LASER LIGHT SCATTERING

CHARLES S. JOHNSON, JR.
M.A. Smith Professor of Chemistry
University of North Carolina

DON A. GABRIEL
Professor of Medicine
University of North Carolina

DOVER PUBLICATIONS, INC.
New York

Bibliographical Note

Laser Light Scattering was originally published in *Spectroscopy in Biochemistry,* Volume II, edited by J. Ellis Bell, copyright 1981. Reprinted with the permission of CRC Press, Boca Raton, Florida. This Dover edition, first published in 1994, is corrected and includes a new Preface.

Library of Congress Cataloging-in-Publication Data

Johnson, Charles S. (Charles Sidney), 1936–
Laser light scattering / Charles S. Johnson, Jr. and Don A. Gabriel.
p. cm.
An unabridged, corrected republication of the work first published as chapter 5 of the book Spectroscopy in biochemistry (volume II), edited by J. Ellis Bell, CRC Press, Boca Raton, Florida, 1981, with a new preface.
Includes bibliographical references (p.).
ISBN 0-486-68328-1
1. Laser spectroscopy. 2. Laser beams—Scattering. 3. Lasers in biochemistry. I. Gabriel, Don A. II. Title.
QP519.9.L37J64 1994
674.19'285—dc20 94-24205
CIP

Manufactured in the United States by Courier Corporation
68328106
www.doverpublications.com

PREFACE

The availability of lasers and the development of dynamic light scattering methods have led to a rebirth of interest in light scattering applications in polymer science and biophysical chemistry. The spectral information in scattered light is now routinely used to measure diffusion coefficients, flow rates, and relaxation time distributions for complex systems.

Our original motivation for writing this brief text was to provide an introduction to modern light scattering methods that would encompass both classical methods and various forms of dynamic light scattering including photon correlation spectroscopy and electrophoretic light scattering with emphasis on biochemical applications. In the ensuing years instrumentation and data analysis methods have improved, but the basic concepts have not changed. Therefore, we are delighted that this book can now be made available to students and beginning research workers in a convenient format and at a modest price.

We hope that this text stimulates interest in scattering methods, and we have provided numerous references for additional reading. In particular we note the fine text by B. J. Bern and R. Pecora (*Dynamic Light Scattering,* Wiley, 1976). Also, we recommend the new introductory texts by Kenneth S. Schmitz (*An Introduction to Scattering by Macromolecules,* Academic Press, 1990) and Ben Chu (*Laser Light Scattering: Basic Principles and Practice,* Academic Press, 1991). For those requiring a comprehensive, up-to-date treatment of modern research applications we recommend *Laser Light Scattering* edited by Wyn Brown (Clarendon Press, 1993).

We thank CRC Press for permitting this text to be reprinted with minor corrections. Also, we thank David M. Johnson for preparing the original figures.

CHARLES S. JOHNSON, JR.
DON A. GABRIEL

Chapel Hill, North Carolina
May 1994

TABLE OF CONTENTS

I. INTRODUCTION

Light scattering has provided an important method for characterizing macromolecules for at least three decades. However, the replacement of conventional light sources by lasers in recent years has qualitatively changed the field and has sparked renewed interest. Through the use of intense, coherent laser light and efficient spectrum analyzers and autocorrelators, experiments in the frequency and time domains can now be used to study molecular motion, e.g., diffusion and flow, and other dynamic processes, as well as the equilibrium properties of solutions. The technology for clarifying samples has also significantly improved. Recirculation through submicron filters in closed loop systems reduces the effects of dust and other contaminants, and the new time domain techniques provide built-in tests for the presence of such particles. These advances make laser light scattering a powerful form of spectroscopy for use in biochemistry.

Classical light scattering studies are concerned with the measurement of the intensity of scattered light as a function of the scattering angle. In addition to this kind of study, laser light sources now permit spectral information to be obtained from the scattered light. The latter type of experiment is often called quasi-elastic light scattering (QLS), and the various forms of the experiment are known as light beating spectroscopy (LBS), intensity fluctuation spectroscopy (IFS), and photon correlation spectroscopy (PCS). Related experiments in laser doppler velocimetry (LDV) now permit very low rates of uniform motion to be measured. A special case of LDV is electrophoretic light scattering (ELS) where mobilities are determined. The aim of this book is to provide an introduction to both the classical and quasi-elastic forms of laser light scattering, which can serve as an introductory text for students and a reference for research workers. The same purpose has been served for prelaser classical light scattering since 1960 by Chapter 5 of Tanford's excellent book, *Physical Chemistry of Macromolecules.*[1] As in that chapter, we emphasize concepts and the kinds of information that can be obtained from the various experiments rather than either presenting a comprehensive survey of the literature or discussing experimental techniques in great detail. For the most part we stick to well-developed techniques such as the measurement of translational and rotational diffusion coefficients. A few specialized applications (such as the study of motility) are also discussed; however, we have not included scattering from internal modes in polymers or gels. Interesting work is going on in the latter areas, but the theory is still in flux.

We attempt to keep the objectives in sight at all times and to avoid distracting complications as much as possible. Accordingly, certain definitions and derivations have been relegated to a set of appendixes. In general, we have opted for simplicity rather than elegance, or even rigor in certain cases, in the hope that a wider audience can be reached. For those who desire more extensive treatments there is an abundance of sources. For classical scattering, one of the latest reviews is that by Timasheff and Townsend, which is concerned with proteins.[2] Also Huglin[3] has edited an extensive volume that emphasizes experimental methods and data handling in the study of polymer solutions. The book by Fabelinskii[4] is a comprehensive, but difficult treatment mainly of prelaser light scattering without special attention to macromolecules. The classical physics of light scattering has been treated in detail by Kerker.[5] The book by Long,[6] while emphasizing Raman scattering, provides a good introduction to Rayleigh scattering, especially polarization and electronic resonance effects. Quasi-elastic scattering, also called dynamic light scattering, has received considerable attention in the literature, and the recent review papers are too numerous to list in full. An encyclopedic review of dynamic light scattering from biopolymers and biocolloids has been provided by Schurr.[7] Pusey and Vaughan have given a much briefer review of the principles of intensity fluctuation spectroscopy,[8] and Carlson[9] has covered applications in molecular biology. Laser velocimetry has been reviewed by Ware.[10,11] An introduction to the theoretical foundations of dynamic light scattering is presented in the book by Berne and Pecora,[12] which is an excellent text for students of physical chemistry. The book by Chu[13] is a good reference for many experimental and theoretical details. For those attempting to understand photon statistics and photon correlation spectroscopy, the monograph edited by Cummins and Pike is invaluable.[14]

II. CLASSICAL LIGHT SCATTERING

A. Scattering Intensity

1. Scattering by an Isolated Dipole and by Gases

Lasers produce collimated, quasi-monochromatic radiation having high intensity. In all but the least expensive lasers, the output is highly polarized. For discussion of the scattering experiment we adopt the coordinate system shown in Figure 1, where the incident laser light propagates in the +y-direction, and the x and y axes define the scattering plane. Only the electric field of the incident light is of interest here, and we assume that the incident light is polarized so that the electric field is in the z-direction. We express the electric field of the incident light as

$$E_z = E_{zo}\cos(k_o y - \omega_o t) \tag{1}$$

where $k_o = 2\pi/\lambda_o$, λ_o is the wavelength in vacuum, and $\omega_o = 2\pi\nu_o$ is the laser frequency. The electric field E_z interacts with electrons in an atom or molecule to induce an electric dipole moment, which oscillates at the angular frequency ω_o. The usual expression for the induced dipole is

$$\underset{\sim}{p} = \underset{\approx}{\alpha} \cdot \underset{\sim}{E} \tag{2}$$

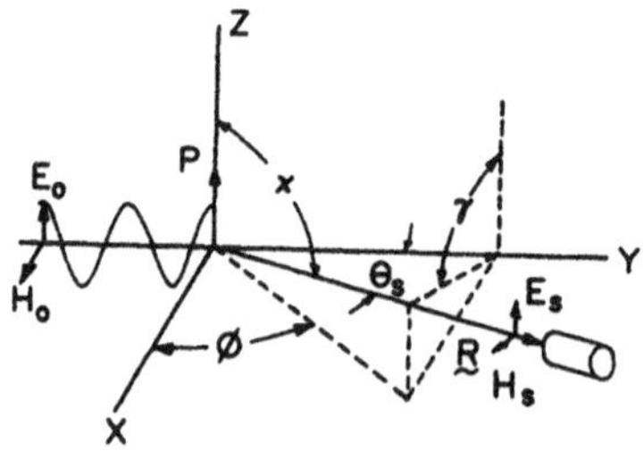

FIGURE 1. Scattering geometry.

where in general p is a vector not necessarily in the same direction as E, and α is the *polarizability* tensor that relates the two vectors. A discussion of the polarizability tensor is given in Appendix A. We shall mainly be concerned with isotropic scatterers for which α is independent of orientation and Equation 2 becomes

$$p = p_z = \alpha E_z = \alpha E_{0z} \cos(k_0 y - \omega_0 t) \tag{3}$$

where α is a constant.

It is well known in electromagnetic theory that an oscillating dipole produces radiation — as for example with the oscillating currents in antennas of radio transmitters. The oscillating electric moment p_z thus provides the source of the scattered light. This is equivalent to the statement that an accelerating charge generates electromagnetic radiation. The solution of Maxwell's equations for an oscillating dipole shows that at large distances, i.e., where $R >> \lambda_o$, the electric field E_s of the scattered light is proportional to $d^2p/dt^2 = -\omega_o^2 p$ and inversely proportional to distance R as required by the conservation of energy. The complete expression in SI units for the field E_s at R resulting from dipole at the origin in Figure 1 is

$$E_s = \frac{-\omega_0^2 p \sin \chi}{(4\pi\epsilon_0) c^2 R} \tag{4}$$

where χ is the angle between the induced moment p and R, ϵ_o is the permittivity of free space, and c is the speed of light. When the detector is in the xy plane, the factor $\sin\chi$ is of course unity.

Detectors, e.g., photomultiplier tubes, respond to the intensity rather than the electric field of incident light. The instantaneous intensity, which is defined as the rate of passage of energy through unit area perpendicular to the direction of propagation, is given by

$$I = c\epsilon_0 E^2 \tag{5}$$

Since all measurements require times that are much larger than the oscillation period of the radiation field, we are usually concerned with the cycle average of I. Thus for the incident radiation

$$\overline{I_i} = c\epsilon_o\overline{E^2} = c\epsilon_o E_{oz}^2 \frac{\omega_o}{2\pi}\left[\int_o^{2\pi/\omega_o} \cos^2 \omega_o t \, dt\right] \tag{6}$$

$$= \frac{c\epsilon_o}{2} E_{oz}^2$$

In calculations involving electric fields it is often convenient to use complex variables, e.g.,

$$E = E_{oz} e^{i(k_o y - \omega_o t)}$$

rather than $E = E_{oz} \cos(k_o y - \omega_o t)$. When the complex variables are handled properly, derivations are simplified, but the final results are not changed. For example, if the electric field of the incident radiation is written in complex form, the cycle average of the intensity can be written as[15]

$$\overline{I_i} = c\epsilon_o \overline{(\text{Re } E)^2} = \frac{c\epsilon_o}{2} E^* E = \frac{c\epsilon_o}{2} |E|^2 \tag{7}$$

where Re means "the real part of" and E* is the complex conjugate of E.

Using Equation 6 the intensity of the scattered light is given by $\overline{I_s} = c\epsilon_o\overline{E^2}$, and Equation 4 yields

$$\overline{I_s} = c\epsilon_o \left[\frac{\omega_o^4 \sin^2 \chi}{(4\pi\epsilon_o)^2 c^4 R^2}\right] \overline{p^2} \tag{8}$$

From Equations 3 and 6 we see that $c\epsilon_o\overline{p^2} = \alpha^2 \overline{I_i}$, and Equation 8 gives

$$\frac{\overline{I_s}}{\overline{I_i}} = \frac{\omega_o^4 \alpha^2 \sin^2 \chi}{(4\pi\epsilon_o)^2 c^4 R^2} = \frac{16\pi^4 \alpha^2 \sin^2 \chi}{(4\pi\epsilon_o)^2 \lambda_o^4 R^2} \tag{9}$$

The inverse fourth power dependence on λ_o was predicted by Lord Rayleigh on the basis of simple dimensional arguments.[16] It accounts for the blue color of the sky, since molecules in the atmosphere tend to scatter the blue part of the solar spectrum with greater intensity than the longer wavelength red components. At optical frequencies almost all of the scattering results from electrons, and the number of electrons increases with molecular volume. Therefore, the polarizability is roughly proportional to molecular volume, especially for larger molecules, and the intensity accordingly depends on the square of the volume. In Equation 9 and in other parts of this book we have used SI units, which are based on the MKS system, except when established definitions were preserved in cgs units. The equations can be converted to the cgs system simply by replacing $4\pi\varepsilon_0$ with unity.

The intensity of scattering per unit volume for a gas at low pressure can be obtained by multiplying Equation 9 by N, the number of scatterers per unit volume. It is conventional, however, to rewrite this equation in terms of the change of refractive index with concentration, which is the experimentally determined quantity, rather than the molecular polarizability. If C is the mass per unit volume of the gas, then at low pressures the refractive index n can be expanded in a Taylor's series

$$n = 1 + \left(\frac{\partial n}{\partial C}\right)_0 C \tag{10}$$

so that

$$n^2 \simeq 1 + 2\left(\frac{\partial n}{\partial C}\right)_0 C \tag{11}$$

However, as discussed in Appendix B, n^2 is also given by

$$n^2 = 1 + N\alpha/\epsilon_0 \tag{12}$$

The combination of Equations 11 and 12 permits us to write

$$\alpha = \frac{2\epsilon_0}{N}\left(\frac{\partial n}{\partial C}\right)_0 C = 2\epsilon_0 m\left(\frac{\partial n}{\partial C}\right)_0 \tag{13}$$

where m is the mass per scattering particle. To obtain the intensity of scattering per unit volume, we substitute Equation 13 into Equation 9 and multiply by $N = N_A C/M$ where N_A is Avogadro's number and M is the molecular weight. Thus

$$\frac{\overline{I}_s}{\overline{I}_i} = \frac{4\pi^2 \sin^2\chi\, CM}{\lambda_0^4 R^2 N_A}\left(\frac{\partial n}{\partial C}\right)_0^2 \tag{14}$$

Scattering experiments are usually reported in terms of the Rayleigh ratio $\mathcal{R}_\theta$ defined by

$$\mathcal{R}_\theta = \left(\frac{I_\theta}{V}\right)\frac{R^2}{I_i} = \frac{I_s}{I_i}R^2 \tag{15}$$

where I_θ is the measured intensity at θ_s, V is the volume of the scattering region, and the bars on I_s and I_i have been dropped, as they will be henceforth, for convenience. Equation 15 is not very useful in practice, since the radiant power P rather than the intensity I is usually measured by photomultiplier tubes and the scattering volume V is not known with accuracy. A more practical expression for $\mathcal{R}_\theta$ is shown in Equation 16.[17]

$$\mathcal{R}_\theta = \left(\frac{P_\theta}{P_i}\right) \frac{1}{(\Delta\Omega)\ell} \qquad (16)$$

Here P_θ is the radiant power of the light collected at the scattering angle θ, P_i is the radiant power of the incident beam, $\Delta\Omega$ is the solid angle of the scattered light that is collected, and ℓ is the length of the scattering volume. The conversion of Equation 15 into Equation 16 proceeds by using the cross-sectional areas A_i and $A_\theta = R^2(\Delta\Omega)$ of the incident and scattered beam, respectively, where R is the distance from the scattering volume to the detector, with the definitions $I_\theta = P_\theta/A_\theta$, $I_i = P_i/A_i$, and $V = \ell A_i$.

One additional comment about Equation 14 is in order before we proceed to the consideration of condensed phases. We have assumed that the incident light is polarized (along the z-direction) and that the scattering particles have dimensions much smaller than the wavelength of the incident light. As a consequence, the intensity is independent of the scattering angle θ_s. However, if either (1) the incident light is unpolarized, or (2) all of the scattered light in the cone θ_s to $\theta_s + d\theta_s$ is collected regardless of the angle χ, then the factor sin χ must be replaced by $(1 + \cos^2\theta_s)/2$.[1] This factor is obtained by averaging $\sin^2\chi$ over the angle γ. By inspection of Figure 1 it is seen that $\cos\chi = \sin\theta_s \cos\gamma$. Therefore, $\sin^2\chi = 1 - \sin^2\theta_s\cos^2\gamma$ and

$$\langle\sin^2\chi\rangle = \frac{1}{2\pi}\int_0^{2\pi} \sin^2\chi \, d\gamma = \frac{1}{2}(1 + \cos^2\theta_s) \qquad (17)$$

In practice we find that either χ is set at 90° or that light is collected at all values of χ. In the following χ is assumed to be 90° unless otherwise stated.

2. Scattering by Macromolecules in Solution

In contrast to light scattered from gases, the intensity of scattered light from condensed phases is less than that predicted by Equations 9 and 14. The reduced intensity is the result of destructive interference. In fact, for perfect crystals irradiated with light, the wavelength of which is much greater than the separation of the lattice planes, no light is scattered. This follows from the fact that it is always possible to pair two scattering planes so that destructive interference occurs. Scattering from crystals is possible at certain angles, however, when the wavelength of the incident radiation is roughly equal to the distance d separating the scattering planes. The condition for scattering is the well-known Bragg relation, $\sin(\theta_s/2) = n\lambda/2d$ where θ_s is defined in Figure 1.

On the other hand, the scattering centers in liquids are not stationary, but undergo Brownian movement, which produces transient optical inhomogenities in the solution.

It is the presence of these inhomogenities that allows a small fraction of the scattered radiation to escape destructive interference and be observed as scattered intensity outside the sample. Our task is to formulate a theory that relates the intensity of scattered light to these transient optical inhomogenities. The basic ideas in the fluctuation theory of light scattering are attributed to Smoluchowski[18] and Einstein.[19] The problem can be approached by considering the solution of scatters as being composed of N small volume elements per unit volume. The volume elements δV are assumed to be small relative to the wavelength of the incident radiation. The light scattered from the independent volume elements largely cancels, but at any instant there will be a deviation from the time average number of particles in any volume element, and the cancellation will be incomplete. The connection with the scattering theory developed in Section II.A.1 is made by realizing that fluctuations in concentration or density lead to fluctuations in the polarizability. The fluctuation in the polarizability of one volume element is defined as $\delta\alpha_v = \alpha_v - \overline{\alpha}_v$ where α_v is the instantaneous polarizability and $\overline{\alpha}_v$ is the time average of α_v. A fluctuation can obviously be either positive or negative, and from the definition the time average of $\delta\alpha_v$ is zero. In Equation 9 the intensity of the scattered radiation is shown to be proportional to the square of the polarizability. To obtain the contribution from an average volume element, we square the quantity, $\alpha_v = \bar{\alpha}_v + \delta\alpha_v$, and take the time average. Thus,

$$\overline{(\overline{\alpha}_V + \delta\alpha_V)^2} = (\overline{\alpha}_V)^2 + \overline{(\delta\alpha_V)^2} \tag{18}$$

The contribution from $(\overline{\alpha}_v)^2$ cancels exactly as in perfect crystals, and the net scattering intensity depends on $\overline{(\delta\alpha_v)^2}$. Substituting this result into Equation 9 and multiplying by $N = 1/\delta V$ gives

$$\frac{I_s}{I_i} = \frac{16\pi^4 \overline{(\delta\alpha_V)^2}}{(4\pi\epsilon_0)^2 \lambda_0^4 R^2 \delta V} \tag{19}$$

for the scattering intensity per unit volume. To obtain $\overline{(\delta\alpha_v)^2}$ in terms of more readily measurable experimental quantities, we note that the mean square fluctuation in the polarization for a given volume element is related to the mean square fluctuation in concentration by

$$(\delta\alpha_V)^2 = \left(\frac{\partial\alpha}{\partial C}\right)^2_{T,V} \overline{(\delta C)^2} \tag{20}$$

The smaller contributions resulting from temperature and volume fluctuations are neglected in this derivation, since these contributions are expected to be the same for the solution as for the solvent. It is expected that the experimental measurements can be corrected so that only the "excess" scattering by the solute is obtained.

In Appendix B it is shown that the polarizability is related to the refractive index of the solution by

$$n^2 - n_0^2 = \frac{\alpha}{(\delta V)\epsilon_0} \tag{21}$$

This equation can be differentiated with respect to the concentration to obtain an expression for $(\partial \alpha/\partial C)_{T,V}$ in terms of the measurable quantity $(\partial n/\partial C)_{T,V}$. Thus

$$2n\left(\frac{\partial n}{\partial C}\right)_{T,V} = \frac{1}{(\delta V)\epsilon_0}\left(\frac{\partial \alpha}{\partial C}\right)_{T,V} \tag{22}$$

The relationship between the mean-square fluctuation in the polarizability and the mean-square fluctuation in the concentration is obtained by combining Equations 20 and 22.

$$\overline{(\delta\alpha_V)^2} = [2n(\delta V)\epsilon_0]^2\left(\frac{\partial n}{\partial C}\right)^2_{T,V}\overline{(\delta C)^2} \tag{23}$$

Equation 23 can now be used with Equation 19 to obtain the intensity of the scattered light in terms of the mean-square concentration fluctuations.

$$\frac{I_s}{I_i} = \frac{4\pi^2\, n^2(\delta V)\,(\partial n/\partial C)^2_{T,V}\,\overline{(\delta C)^2}}{\lambda_0^4\, R^2} \tag{24}$$

B. Concentration Dependence

The magnitude of the average concentration fluctuation will clearly depend on the energy required to produce the fluctuation. A simple calculation that is exhibited below shows that $\overline{\delta C^2}$ depends on $(\partial^2 A/\partial C^2)_{T,V}$ where A is the Helmholtz free energy. Again we imagine that the sample is divided into small volume elements δV, which exchange solute particles and are in thermal equilibrium. The problem then becomes one of determining the probability of fluctuations in the composition of a given volume element, and determining the mean-square concentration fluctuation. By analogy with chemical kinetics the probability of a fluctuation $(\delta C)^2$ is given by $e^{-\delta A/k_BT}$ where δA is the associated increment in the Helmholtz free energy for the volume element. Since the fluctuations are small, δA can be expanded in a Taylor's series

$$\delta A = \left(\frac{\partial A}{\partial C}\right)_{T,V}\delta C + \frac{1}{2!}\left(\frac{\partial^2 A}{\partial C^2}\right)_{T,V}(\delta C)^2 + \ldots \tag{25}$$

The first term on the right vanishes, since the system is at equilibrium. Therefore, the probability of a fluctuation becomes

$$\exp(-\delta A/k_BT) = \exp\left[-\left(\frac{\partial^2 A}{\partial C^2}\right)_{T,V}(\delta C)^2/2k_BT\right] \tag{26}$$

This weighting factor can now be used to calculate $\overline{(\delta C)^2}$ as follows:

$$\overline{(\delta C)^2} = \frac{\int_0^\infty (\delta C)^2 \, e^{-\delta A/k_B T} \, d(\delta C)}{\int_0^\infty e^{-\delta A/k_B T} \, d(\delta C)} \tag{27}$$

The above integrals have the standard form $\int_0^\infty x^n e^{-ax^2} dx$ and can be found in tables. Therefore, the evaluation of Equation 27 immediately gives

$$\overline{(\delta C)^2} = k_B T/(\partial^2 A/\partial C^2)_{T,V} \tag{28}$$

The quantity $(\partial^2 A/\partial C^2)_{T,V}$ depends on the concentration, and thus Equation 28 in conjunction with Equation 24 gives the concentration dependence of the scattered intensity. A series expansion of $(\partial^2 A/\partial C^2)_{T,V}$ in terms of the concentration could be introduced at this point. However, it is conventional to relate $(\partial^2 A/\partial C^2)_{T,V}$ to $(\partial \mu_1/\partial C)_{T,V}$ so that the concentration dependence is given in terms of the familiar virial coefficients. The derivation of this relationship is given in Appendix C with the important result that

$$\left(\frac{\partial^2 A}{\partial C^2}\right)_{T,V} = -\frac{\delta V}{C\overline{V}_1}\left(\frac{\partial \mu_1}{\partial C}\right)_{T,V} \tag{29}$$

where $\overline{V}_1$ and μ_1 are the partial molar volume and the chemical potential of the solvent, respectively. The virial expansion of $(\partial \mu_1/\partial C)_{T,V}$, which is also discussed in Appendix C, is given by

$$-\frac{1}{k_B T \overline{V}_1}\left(\frac{\partial \mu_1}{\partial C}\right)_{V,T} = N_A\left[\frac{1}{M} + 2B_2 C + 3B_3 C^2 + \ldots\right] \tag{30}$$

where M is the molecular weight of the solute and the B_n are virial coefficients. Equations 28 to 30 can then be combined with Equation 24 to yield

$$\frac{I_s}{I_i} = \frac{4\pi^2 n^2 C(\partial n/\partial C)_0^2}{\lambda_0^4 R^2 N_A [M^{-1} + 2B_2 C + 3B_3 C^2 + \ldots]} \tag{31}$$

Equation 31 describes the intensity of the "excess scattered light" per unit volume from a dilute solution of scatterers. The term "excess scattering" is used to describe the scattering resulting from the solute alone. In fact scattering from the solvent, which is usually a weak aqueous salt solution, is subtracted from the total scattered intensity in the data analysis as a blank correction. As discussed in Section III.A, Equation 31 has been derived for polarized incident light. Therefore, the $\sin^2\chi$ factor must be replaced by $(1 + \cos^2\theta_s)/2$ for instruments using unpolarized light, or when all the scattered light in the angular range θ_s to $\theta_s + d\theta_s$ is collected.

Equations 15 and 31 can be combined to obtain

$$\frac{\mathcal{K}c}{\mathcal{R}_\theta} = \frac{1}{M} + 2B_2C + 3B_3C^2 + \ldots \tag{32}$$

where $\mathcal{K}$, the optical constant, is defined by

$$\mathcal{K} = \frac{4\pi^2 n^2 (\partial n/\partial C)_0^2}{\lambda_0^4 N_A} \tag{33}$$

It should be noted that the optical constant defined in Equation 33 is twice as large as that used by Tanford.[1] This equation has been derived for small particles that conform to the rule that their major dimension is less than $\lambda/10$. It is obvious that the intercept of a plot of $\mathcal{K}\,C/\mathcal{R}_\theta$ vs. C gives the inverse molecular weight and the initial slope is $2B_2$, where B_2 is the second virial coefficient. As discussed in Appendix D, the molecular weight obtained when there is polydispersity is the weight average molecular weight. In the limit of $C = 0$ Equation 31 becomes equivalent to Equation 14. The purpose of the simple derivation outlined here is to obtain the concentration dependence of the scattered intensity for a real solution of macromolecules.

Light scattering intensities are sometimes reported in terms of the turbidity, i.e., the attenuation coefficient τ for a beam passing through the sample. Attenuation, of course, results from scattering in any direction except $\theta_s = 0$. For small particles, equations relating R_θ and τ can easily be derived with the help of Equations 15 and 17. For light passing through the volume $V = A_i\,\Delta\ell$, the average intensity scattered into the cone defined by θ_s and $\theta_s + d\theta_s$ is

$$I_\theta = \frac{\mathcal{K}CM}{2R^2}\,[1 + \cos^2\theta_s]\,I_i\,A_i\,\Delta\ell \tag{34}$$

The power scattered into this angular range is $dP_\theta = I_\theta\,2\pi\,R^2\sin\theta_s\,d\theta_s$ and the total power lost from the incident beam in the path length $\Delta\ell$ is obtained by integrating over θ_s. Thus

$$\begin{aligned}\Delta P &= -\int dP_\theta \\ &= -\mathcal{K}CM\,P_i\,\Delta\ell\,\pi\int_0^\pi [1 + \cos^2\theta_s]\sin\theta_s\,d\theta_s \\ &= -\mathcal{K}CM\,P_i\left(\frac{8\pi}{3}\right)\Delta\ell\end{aligned} \tag{35}$$

For an infinitesimal path length Equation 35 has the form $dP/P = -\tau d\ell$, which integrates to give $P = P_i \exp(-\tau\ell)$. Equation 35 shows that the attenuation coefficient is just $\tau = (8\pi/3)\mathcal{K}$ CM. Another optical constant $H = (8\pi/3)\mathcal{K}$ is often introduced so that

$$\frac{HC}{\tau} = \left(\frac{\mathcal{K}C}{\mathcal{R}_\theta}\right)_{C=O} \tag{36}$$

This equation becomes equivalent to Equation 32 when the concentration dependence is introduced.

C. Size Dependence

1. Structure Factors

It is evident from Figure 2 that when the size of the scattering particle increases to the point where the largest dimension is greater than $\lambda/10$, the path difference for light scattered from two points in the particle may become large enough to produce a significant phase difference and hence interference in the scattered light. The interference will obviously attenuate the intensity of the scattered light. Initially, the decrease in intensity may seem a nuisance, but it turns out to be a unique source of information pertaining to the size and shape of the particle. The dependence of the scattered intensity on the particle size and shape is contained in the structure factor $S(\underset{\sim}{K})$, which is defined by

$$S(\underset{\sim}{K}) = \frac{\text{scattered intensity observed (at } \theta_s)}{\text{scattered intensity without interference (at } \theta_s = O)} \tag{37}$$

In our development of the expression for $S(\underset{\sim}{K})$ we assume randomly oriented, nonabsorbing particles, composed of optically isotropic material. The refractive index within the particle is assumed to be homogeneous, and the refractive indexes of the particle and solvent are assumed to be only slightly different so that the Rayleigh-Debye condition, $4\pi L(\Delta n)/\lambda << 1$, is satisfied[5] where L is the largest dimension of the particle and Δn is the difference between the refractive indexes of the particle and the solvent.

Figure 2 shows a macromolecule with a scattering segment S at position $\underset{\sim}{\rho}_j$ relative to the center of mass. Parallel incident light with the wave vector $\underset{\sim}{k}_i$ passes through reference plane OO′. Similarly the scattered light passes through reference plane PP′. The path difference between light scattered from segment S and that passing through the origin at O is given by

$$\text{path difference} = O' S P' - OP$$

The phase difference ϕ for the two rays is just $2\pi n/\lambda_o$ times the path difference. Therefore, we can write

$$\phi = \underset{\sim}{\rho}_j \cdot \underset{\sim}{k}_i - \underset{\sim}{\rho}_j \cdot \underset{\sim}{k}_s = \underset{\sim}{\rho}_j \cdot \underset{\sim}{K} \tag{38}$$

where $\underset{\sim}{K} = \underset{\sim}{k}_i - \underset{\sim}{k}_s$ is called the scattering vector. Simple trigonometry shows that

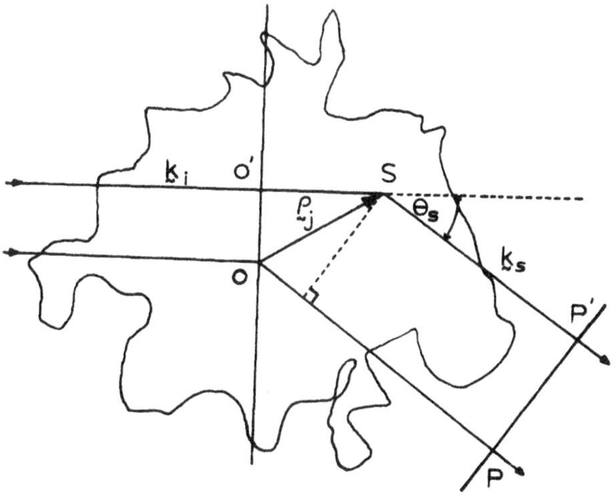

FIGURE 2. Diagram of the path difference for scattering from a segment in a macromolecule.

$$|\underset{\sim}{K}| = \frac{4\pi n}{\lambda_o} \sin\left(\frac{\theta_s}{2}\right) \tag{39}$$

where we have used the fact that $|\underset{\sim}{k}_s| = |\underset{\sim}{k}_i| = (2\pi n/\lambda_o)$. Accordingly, the electric field of the light scattered by the jth segment can be written as

$$E_j = a_j e^{i\underset{\sim}{K} \cdot \underset{\sim}{\rho}_j} E_o e^{-i\omega_o t} \tag{40}$$

where a_j contains the factors such as the polarizability, the distance R to the detector, etc. To calculate the intensity of the scattered light from the total particle for a given scattering angle, the electric fields of the scattered waves from the P segments within the particle must be summed as follows:

$$\sum_{j=1}^{P} E_j = E_o e^{-i\omega_o t} \sum_{j=1}^{P} a_j e^{i\underset{\sim}{K} \cdot \underset{\sim}{\rho}_j} \tag{41}$$

According to Equation 7 the intensity of the scattered light is

$$I_s = \frac{c\epsilon_o}{2} \left| \sum_{j=1}^{P} a_j e^{i\underset{\sim}{K} \cdot \underset{\sim}{\rho}_j} \right|^2 E_o^2 \tag{42}$$

Equation 42 gives the scattering for one particle with a fixed orientation. For an ensemble of randomly oriented particles the intensity is given by

$$I_s = N \frac{c\epsilon_o}{2} < \left| \sum_{j=1}^{P} a_j e^{i\underset{\sim}{K} \cdot \underset{\sim}{\rho}_j} \right|^2 >_{angle} E_o^2 \tag{43}$$

where N is the number of particles per unit volume. The calculation of the structure factor also requires the intensity in the absence of interference. This condition is met at $\theta_s = O$, where the path distance for all scattering segments becomes identical; therefore, the intensity without interference is given by

$$<I_s>_{\theta=O} = \frac{c\epsilon_o}{2} | a |^2 P^2 E_o^2 N \tag{44}$$

where all of the a_j have been assumed to be equal, i.e., $a_j = a$ for all j. The expression for the structure factor then follows from Equation 37.

$$S(\underset{\sim}{K}) = < \left| \frac{1}{P} \sum_{j=1}^{P} e^{i\underset{\sim}{K} \cdot \underset{\sim}{\rho}_j} \right|^2 >_{angle} \tag{45}$$

To illustrate this procedure, we consider two simple examples:

a. ***Long Thin Rod***

The rod is considered as a linear array of uniformly polarizable segments, and the cross section is assumed to be sufficiently small that interference can only arise for scattering from different positions along the major axis. For a uniform rod the summation in Equation 45 can be replaced with an integral over the length L

$$\lim_{P\to\infty} \sum_{j=1}^{P} \frac{1}{P} e^{i\underset{\sim}{K} \cdot \underset{\sim}{\rho}_j} = \frac{1}{L} \int_{-L/2}^{L/2} e^{iK\rho \cos \alpha} \, d\rho = \frac{\sin w}{w} \tag{46}$$

where $w = (KL/2) \cos \alpha$ and α is the angle formed by the scattering vector $\underset{\sim}{K}$ and the position vector $\underset{\sim}{\rho}_j$ as shown in Figure 3. This result can now be substituted into Equation 45 and the averaging carried out over all orientations. We proceed by considering the probability of finding an orientation in the solid angle between Ω and $\Omega + d\Omega$. The integration over the solid angle Ω is carried out using the relations given in Figure 3.

$$S(\underset{\sim}{K}) = \frac{1}{4\pi} \int_{\phi=O}^{2\pi} \int_{\alpha=0}^{\pi} \left| \frac{\sin w}{w} \right|^2 \sin \alpha \, d\alpha d\phi \tag{47}$$

Equation 47 is then evaluated by standard methods utilizing the trigonometric identity $2 \sin^2 \theta = (1 - \cos 2\theta)$ and integration by parts. The result is[12]

$$S(\underset{\sim}{K}) = \frac{2}{KL}\int_{0}^{KL} \frac{\sin x}{x}\, dx - \left[\frac{\sin(KL/2)}{(KL/2)}\right]^2 \qquad (48)$$

This equation was first given by Neugebauer[20] in 1943. The first term is usually evaluated by series expansion when KL is small. For large values of KL, restrictions imposed by the Rayleigh-Debye condition may be a limiting factor in some experimental situations.

b. Uniform Sphere

In order to calculate the structure factor for a sphere of radius R, we again return to Equation 45. This time the summation is replaced by an integration over the volume of the sphere. In spherical polar coordinates this gives

$$S(\underset{\sim}{K}) = \left| 2\pi \int_{\rho=0}^{R}\int_{\alpha=0}^{\pi} e^{iK\rho\cos\alpha} \sin\alpha d\alpha\, \rho^2 d\rho \Big/ 4\pi \int_{\rho=0}^{R} \rho^2 d\rho \right|^2$$

$$= \left\{ \frac{3}{R^3} \int_{\rho=0}^{R} \frac{\sin(K\rho)}{K\rho}\, \rho^2 d\rho \right\}^2 \qquad (49)$$

Integration by parts can be applied in the last step to obtain

$$S(\underset{\sim}{K}) = \left\{ \frac{3}{X^3} (\sin X - X\cos X) \right\}^2 \qquad (50)$$

where $X = KR$. It is important to note that integration over all orientations of the particle is unnecessary in this case because of the spherical symmetry.

The preceding two examples serve to illustrate the principles involved in the derivation of the structure factor. In practice it is often convenient to carry out the average over orientations in Equation 45 before applying the equation to a particular particle shape. The procedure is as follows. We first rewrite Equation 45 as

$$S(\underset{\sim}{K}) = \frac{1}{P^2} \left\langle \sum_{i=1}^{P}\sum_{j=1}^{P} e^{i\underset{\sim}{K}\cdot\underset{\sim}{r}_{ij}} \right\rangle_{angle} \qquad (51)$$

where $r_{ij} = \underset{\sim}{\rho}_j - \underset{\sim}{\rho}_j$ Using $\underset{\sim}{K} \cdot \underset{\sim}{r}_{ij} = K r_{ij} \cos\alpha$ and the relations in Figure 3 we find that

$$\langle e^{iKr_{ij}\cos\alpha} \rangle_{angle} = \frac{1}{2}\int_{0}^{\pi} e^{iKr_{ij}\cos\alpha} \sin\alpha d\alpha$$

$$= \frac{\sin(Kr_{ij})}{Kr_{ij}} \qquad (52)$$

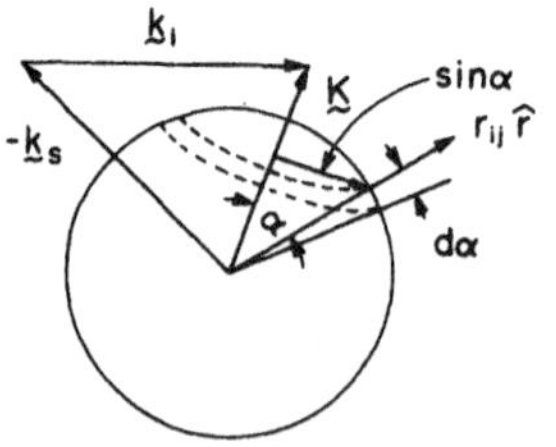

FIGURE 3. The possible orientations of $\underset{\sim}{r}_{ij}$ relative to the scattering vector $\underset{\sim}{K}$. Unit sphere: area in strip (α to $\alpha + d\alpha$) = $2\pi \sin\alpha d\alpha$. Total surface area = 4π.

Equations 51 and 52 combine to give the standard form of the structure factor equation

$$S(\underset{\sim}{K}) = \frac{1}{P^2} \sum_{i=1}^{P} \sum_{j=1}^{P} \frac{\sin(Kr_{ij})}{Kr_{ij}} \tag{53}$$

This is a very useful equation for deriving expressions relating the angular dependence of the scattered intensity to the shapes of particles. The application of Equation 53 to specific shapes has been discussed in many reviews.[1-5,12] The results for some standard particles are listed in Table 1.

In Figure 4 plots of the inverse structure factor S^{-1} ($\underset{\sim}{K}$) vs. $R^2_G \sin^2(\theta_s/2)$ for several common macromolecular shapes are given. One should be extremely cautious, however, in applying these plots to experimental data since the morphology of the experimental curves for a given shape may be distorted by many factors, e.g., polydispersity and optical anisotropy.[21] The resulting ambiguities seriously reduce the usefulness of Equation 53 for many real systems.

2. Radius of Gyration

Most experimental determinations of molecular dimensions require some preliminary assumption about the shape of the molecule. In contrast, light scattering measurements permit the radius of gyration to be determined from the structure factor $S(\underset{\sim}{K})$ without any assumptions about the shape, provided the measurements can be made at sufficiently low angles.[22] To see how this property of the structure factor comes about, we expand $S(\underset{\sim}{K})$ in a Taylor's series so that

$$S(\underset{\sim}{K}) = \frac{1}{P^2} \sum_{i}^{P} \sum_{j}^{P} \left(1 - \frac{K^2 r^2_{ij}}{3!} + \frac{K^4 r^4_{ij}}{5!} + \ldots\right) \tag{54}$$

For small values of K only the first two terms in the expansion are significant, and since

Table 1

STRUCTURE FACTORS S($\underset{\sim}{K}$) AND RADII OF GYRATION FOR VARIOUS PARTICLES

Particle	S($\underset{\sim}{K}$)	R_G^2
Thin rod	$\frac{2}{x}\int_0^x \frac{\sin u}{u}du - \left[\frac{\sin(x/2)}{(x/2)}\right]^2$; x = KL ; L = length	$\frac{L^2}{12}$
Uniform sphere	$\left[\frac{3}{x^3}(\sin x - x\cos x)\right]^2$; x = KR ; R = radius	$\frac{3R^2}{5}$
Gaussian coil[a]	$\frac{2}{x^2}(e^{-x} - 1 + x)$; $x = K^2\frac{\langle h^2\rangle}{6}$ $\langle h^2\rangle$ = mean square end-to-end distance	$\frac{\langle h^2\rangle}{6}$

[a]Reference 12, p. 168

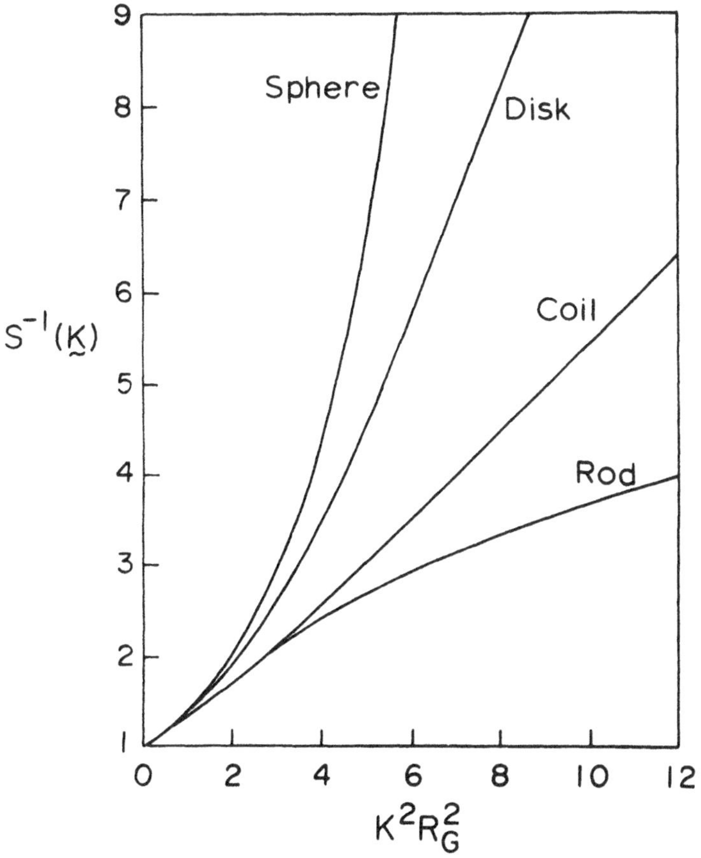

FIGURE 4. The inverse structure factor $S^{-1}(\underset{\sim}{K})$ vs. $K^2R_G^2$ for various shapes (see text).

$$\sum_i^P \sum_j^P 1 = P^2$$

we obtain

$$S(\underset{\sim}{K}) \cong 1 - \frac{K^2}{3!P^2} \sum_i^P \sum_j^P r_{ij}^2 \tag{55}$$

We can introduce the radius of gyration R_G at this point since one of the definitions of this quantity is[23]

$$R_G^2 = \frac{1}{2P^2} \sum_i^P \sum_j^P r_{ij}^2 \tag{56}$$

The use of this definition with Equations 55 and 38 yields the more familiar equation

$$S(\underset{\sim}{K}) = 1 - \frac{16\pi^2 n^2}{3\lambda_0^2} (R_G^2) \sin^2\left(\frac{\theta_s}{2}\right) + \ldots \tag{57}$$

An expansion for the inverse structure factor $S^{-1}(\underset{\sim}{K})$ can easily be obtained from Equation 57 by noting that for small values of x, $(1 - x)^{-1} \simeq 1 + x$. With this factor, Equation 32 can be corrected for interference effects in large particles to give in the limit of zero concentration

$$\left(\frac{\mathcal{K}c}{\mathcal{R}_\theta}\right)_{C=O} = \frac{1}{M_w}\left[1 + \frac{16\pi^2 n^2}{3\lambda_0^2} (R_G^2) \sin^2\left(\frac{\theta_s}{2}\right) + \ldots\right] \tag{58}$$

3. *Zimm Plots*

In a classic article, Zimm[24] reported a graphical technique for simultaneously extrapolating light scattering data to both zero angle and zero concentrations. This is achieved by plotting $\mathcal{K}\,C/\mathcal{R}_\theta$ vs. $\sin^2(\theta_s/2)$ + (constant) C as shown in Figure 5 where the arbitrary constant is chosen to give a convenient spacing of the data points on the graph. It is seen that the data points fall on a grid, producing two families of curves, one corresponding to constant concentration and the other to constant angle. The data points on the grid corresponding to a given angle are then extrapolated to zero concentration, and similarly the points at a given concentration are extrapolated to zero angle. This double extrapolation then produces two new sets of data points, one at zero angle and the other at zero concentration from which the information about the particle shape and weight is obtained. The inverse of the weight average molecular weight is obtained from the intercept of the $\theta = 0$ curve, according to Equation 32, and the second virial coefficient, B_2, is obtained from the slope ($2B_2$ = slope). The intercept of the C = O curve again gives the inverse of the molecular weight, and the initial slope is proportional to the radius of gyration as described by Equation 58. These quantities can be combined to yield

$$R_G^2 = \frac{3\lambda_0^2}{16\pi^2 n^2}\left(\frac{\text{slope}}{\text{intercept}}\right) \tag{59}$$

The radius of gyration is obtained as a z-average (see Appendix D).

There are limitations on the size of the radius of gyration that can be measured by this method. For small values of R_G, the derivation of the structure factor from unity may be so small that an accurate slope cannot be obtained, even when the scattering angle is large. On the other hand, for large values of R_G, it may be impossible to obtain the limiting condition required for Equation 58 to hold. Rough estimates for the limits of R_G using 5° and 160° as the extremes for conveniently obtainable scattering angles and λ_0 = 633 nm as the wavelength give 300 Å and 5000 Å. It is possible to extend these limits by changing the wavelength of the incident radiation. However, with most biological polymers the ultraviolet region is unsuitable because of strong absorption bands in that region, and also in some cases because of photo-induced chemical modification of the polymer. At the opposite extreme where the dimension of the molecule exceeds the wavelength of the incident light, the above theory breaks down and the scattered intensity becomes a complicated oscillatory function of angle. Kerker has given a detailed discussion of this problem.[5] An alternative for large particles is to increase the wavelength of the incident radiation. With the advent of infrared laser sources, e.g., Nd-Yag lasers, dye lasers, and infrared diodes (see Section III.H), as well as improved infrared detectors, large particles can be studied using Rayleigh-Debye theory and Zimm plots. Recently, Morris[25] et al. reported an infrared Zimm plot of *Serratia marcescens,* a bacterium approximately 1 μm in length. CO_2 lasers may offer the possibility of studying particles having diameters of the order of 10 μm.

D. Polydispersity

Inhomogeneity in either the size or shape of the particles leads to ambiguity in the interpretation of the angular dependence of the scattered intensity. Since the measured structure factor is an average over the structure factors of the scattering particles, it is often impossible to characterize polydisperse samples with the limited amount of information available. However, if the shapes of the particles are uniform and known, certain features of the distribution can be assessed. For example a plot of $(K\,C/\mathcal{R}_\theta)$ vs. $\sin^2(\theta/2)$ can yield M_w, M_n, $\langle R^2_G\rangle_z$ and $\langle R^2_G\rangle_n$ for a collection of Gaussian coils. The basic ideas were presented by Zimm[24] who showed that the scattered intensity is proportional to the z-average of the structure factor so that

$$\left(\frac{\mathcal{R}_\theta}{Kc}\right)_{C=O} = M_w \langle S(\underset{\sim}{K})\rangle_z \tag{60}$$

The consequences of this result are discussed in Appendix D.

If the particles are composed of subunits, Equation 60 can be written as

$$\left(\frac{\mathcal{R}_\theta}{Kc}\right)_{C=O} = M_w \frac{\int S(\underset{\sim}{K})_N \, Nf(N)\, dN}{\int Nf(N)\, dN} \tag{61}$$

where $S(\underset{\sim}{K})_N$ is the structure factor for a particle containing N subunits, N is the degree of polymerization, and f(N)dN is the fractional mass in the range N to N + dN. The following definitions are useful:

$$1 = \int f(N)\, dN \tag{62}$$

$$N_n = 1 \Big/ \int \frac{f(N)}{N}\, dN \tag{63}$$

$$N_w = \langle N \rangle = \int N f(N)\, dN \tag{64}$$

$$\langle (\Delta N)^2 \rangle = \int (N - \langle N \rangle)^2 f(N)\, dN = \langle N^2 \rangle - \langle N \rangle^2 \tag{65}$$

The quantity $\langle (\Delta N)^2 \rangle$ is known as the dispersion of the distribution.

In the limit of small scattering angles the series expansion for the structure factor from Equation 57 can be obtained with Equation 61 to give

$$\left(\frac{\mathcal{R}_\theta}{KC}\right)_{C=0} = M_w \left[1 - \frac{16\pi^2 n^2}{3\lambda_0^2 N_w} \int N f(N)\, R_G^2\, dN \sin^2 \left(\frac{\theta_s}{2}\right) \right] \tag{66}$$

To proceed further, a particular particle shape must be assumed. For example, for thin rods $R^2_G = L^2/12$, and we assume that $L = Nb$ where b is the length of a subunit. Therefore

$$\int N f(N)\, R_G^2\, dN = \frac{b^2}{12} \int N^3 f(N)\, dN = \frac{b^2}{12} \langle N^3 \rangle \tag{67}$$

In the case of linear Gaussian coils, Table 1 gives $R^2_G = \langle h^2 \rangle/6$ and $\langle h^2 \rangle$ can be written as Nb^2. Therefore, the required integral becomes

$$\int N f(N)\, R_G^2\, dN = \frac{b^2}{6} \int N^2 f(N)\, dN = \frac{b^2}{6} \langle N^2 \rangle \tag{68}$$

The following discussion considers the case of the Guassian coil. By combining Equations 66 and 68 and taking the inverse we obtain

$$\left(\frac{KC}{R_\theta}\right)_{C=O} = \frac{1}{M_w}\left[1 + \frac{8\pi^2 n^2 b^2}{9\lambda_0^2 N_w} <N^2> \sin^2\left(\frac{\theta_s}{2}\right)\right] \tag{69}$$

This equation can be put in standard form by using Equation 65 to give

$$\left(\frac{KC}{R_\theta}\right)_{C=O} = \frac{1}{M_w}\left\{1 + N_w\left[1 + \frac{<(\Delta N)^2>}{N_w^2}\right]\frac{8\pi^2 n^2 b^2}{9\lambda_0^2}\sin^2\left(\frac{\theta_s}{2}\right)\right\} \tag{70}$$

Equation 66 and subsequent equations all show that the weight average molecular weight and the z-average radius of gyration can be obtained from a plot of $(K\, C/R_\theta)_{C=O}$ vs. $\sin^2(\theta_s/2)$. Equation 70 is important because it relates the slope of the plot to the dispersion of the distribution.

Additional information can be obtained from the slope and intercept of the asymptote of the curve $(K\, C/R_\theta)_{C=O}$ vs. $\sin^2(\theta_s/2)$.[26-28] For example, with the Guassian coil the structure factor listed in Table 1 clearly approaches the asymptotic limit:

$$\lim_{x\to\infty} S(\underset{\sim}{K}) = \frac{2}{x} - \frac{2}{x^2} \tag{71}$$

where $x = K^2Nb^2/6$. For a collection of polydisperse Gaussian coils the expression for the z-average $S(\underset{\sim}{K})$ from Equation 61 an be used to obtain the z-average of Equation 71. Thus

$$\begin{aligned} <S(\underset{\sim}{K})>_z &= \frac{2}{N_w u}\int f(N)\, dN - \frac{2}{N_w u^2}\int \frac{f(N)}{N}\, dN \\ &= \frac{2}{N_w u}\left[1 - \frac{1}{N_n u}\right] \end{aligned} \tag{72}$$

where $u = K^2b^2/6$, and in the last step we have used Equation 63 to introduce the number average of the degree of polymerization N_n. Using the fact that $N_n u$ is large, the inverse of Equation 72 can be written as

$$<S(\underset{\sim}{K})>_z^{-1} = \frac{N_w}{2N_n} + \frac{N_w u}{2} \tag{73}$$

Equation 73 describes a straight line when $<S(\underset{\sim}{K})>_z^{-1}$ is plotted vs. $\sin^2(\theta_s/z)$. This expression can now be combined with Equation 60 to obtain the asymptotic equation

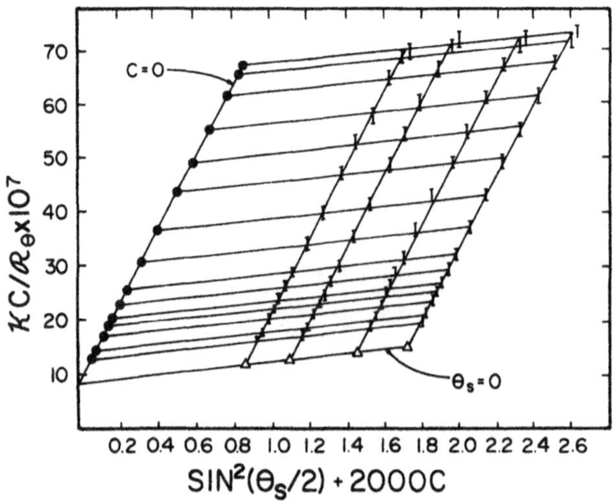

FIGURE 5. Zimm plot for cellulose nitrate. (Reprinted with permission from Benoit, H., Holtzer, A. M., and Doty, P., *J. Phys. Chem.*, 58, 635, 1954. Copyright by the American Chemical Society.)

$$\left(\frac{Kc}{R_\theta}\right)_{C=0} = \frac{N_w}{2M_wN_n} + \frac{N_w}{2M_w}u = \frac{1}{2M_n} + \frac{N_n}{2M_n}u$$

$$= \frac{1}{2M_n} + \frac{(N_nb^2)}{12M_n}\left(\frac{4\pi n}{\lambda_o}\right)^2 \sin^2(\theta_s/2) \qquad (74)$$

If M_o is the molecular weight of a subunit, then $M_w = M_oN_w$ and $M_n = M_oN_n$. These relations were used in deriving Equation 74. The conclusion is that the intercept of the asymptote gives $1/(2M_n)$ and the slope is proportional to N_nb^2, which is equal to the number average of the end-to-end distance. The features of this type of plot are illustrated in Figure 6.

A similar analysis has been presented by Holtzer[29] and Goldstein[30] for rigid rods. Also, Rice[31] and Benoit[26] have considered polydispersity of shape. In spite of the intriguing possibilities suggested by these analyses, it should be realized that the asymptote may not be attainable in many experimental situations.[32] Carpenter[33] has shown that multiple wavelengths of incident light may be used in some situations in order to achieve the asymptotic condition.

III. DYNAMIC LIGHT SCATTERING

A. Time and Frequency Dependence

1. Background

In Section II we were concerned with the average intensity of the scattered light and its dependence on both the concentration of scattering particles and the scattering angle. It was concluded that in favorable cases both the weight average molecular weight and the z-average radius of gyration could be determined from such experiments. In this section we consider the use of light scattering to determine the rates of motion of macromolecules. Motional effects are evident in many kinds of spectroscopies, one of the best known examples being the Doppler broadening, which is encountered in gas

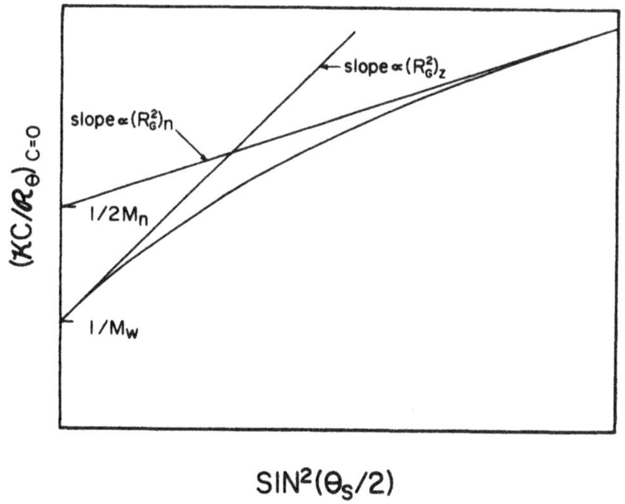

FIGURE 6. Plot showing limiting behavior for a polydisperse system of Gaussian coils. (Reprinted with permission from Benoit, H., Holtzer, A. M., and Doty, P., *J. Phys. Chem.*, 58, 635, 1954. Copyright by the American Chemical Society.)

Table 2
PERFORMANCE FIGURES FOR SPECTROMETERS FOR INCIDENT LIGHT HAVING THE FREQUENCY 5×10^{-14} Hz

Instrument	Resolution	Minimum line width (Hz)	Time scale
Grating spectrograph	10^5	3×10^9	300—0.1 psec
Fabry-Perot interferometer	10^8	5×10^6	200 nsec—30 psec
Lightbeating spectrometer	$\sim 10^{14}$	~ 1	1 sec—1 μsec

phase absorption spectroscopy. A consideration of the velocities of macromolecules, both in diffusive motion and in directed flow, such as that produced in electrophoresis, shows that linewidths and shifts in the range from 1 to 10^5 Hz are expected. Somewhat higher frequencies are found if rotational motions are considered.

To put the frequency shifts in context, one should recall the capabilities of the available types of spectroscopic instruments. In Table 2 we list typical performance figures for grating spectrographs, interferometers, and lightbeating spectrometers for a spectral range in the vicinity of $5 \times 10^{+14}$ Hz, i.e., in the middle of the visible spectrum. Roughly speaking, the resolution is defined as the spectral frequency of a line divided by the contribution to the linewidth, which results from instrumental limitations alone. It is clear from the figures given that the measurement of linewidths in the range 1 to 10^4 Hz requires several orders of magnitude higher resolution than can be obtained by standard optical techniques. In fact, this frequency range was totally inaccessible prior to the development of the light beating experiment in 1964.[34] This remarkable improvement in resolution results from an analysis of the fluctuations in the intensity of the scattered light rather than an attempt to disperse the light according to wavelength. As we shall show, by dealing with the time dependence of the intensity we effectively

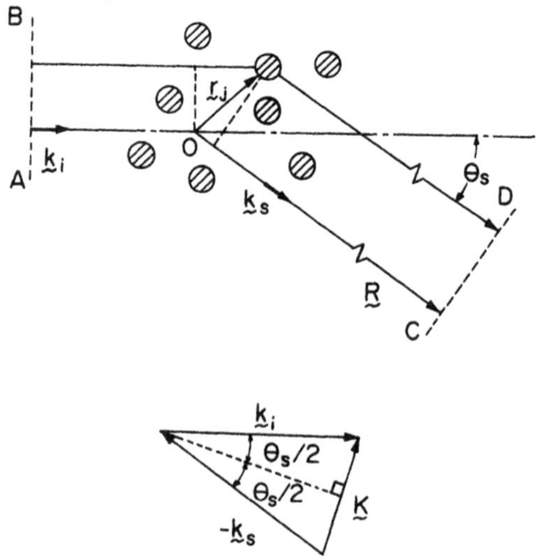

FIGURE 7. The scattering wave vector $\underset{\sim}{K}$ for a solution of small particles.

shift the spectral line down to zero frequency where electronic means can be used for the spectral analysis.

2. The Intensity

In considering the time and frequency dependence of the intensity, we choose to express the electric field in complex form. This is done only for convenience in handling the mathematics. The relationship between the cycle average intensity and the complex electric field is given in Equation 7 as $I = (c\varepsilon_o/2)|E|^2$. Since we are concerned only with the relative intensity, we suppress the factor of $c\varepsilon_o/2$ and write

$$I = EE^* = |E|^2 \tag{75}$$

In spite of the cycle average, I_s will still have time dependence resulting from fluctuations on a longer time scale. To see how the fluctuations in I_s might arise, consider the situation in Figure 7 where incident light having the wave vector $\underset{\sim}{k}_i$ is scattered at the angle θ_s by small particles in suspension. Neglecting the scattering by solvent molecules, the electric field of the scattered light is a superposition of contributions from the N solute particles in the scattering volume.

$$E_s = \sum_{i=1}^{N} E_i \tag{76}$$

The line AB in Figure 7 represents a plane of constant phase for the incident light, and all of the scattering particles are assumed to be coherently illuminated, i.e., to lie within one coherence length along the beam as specified in Section III.H.1. Also, the structure factor $S(\underset{\sim}{K})$ for each particle is assumed to be unity. Interference will still occur at the plane CD in the far field limit because of the different path lengths traversed by light scattered from different particles. Relative to the ray path that passes through the reference point O, the field E_j at CD associated with the jth particle will have the phase

$$\phi_j = \underset{\sim}{k}_i \cdot \underset{\sim}{r}_j - \underset{\sim}{k}_s \cdot \underset{\sim}{r}_j$$
$$= \underset{\sim}{K} \cdot \underset{\sim}{r}_j$$

where we have defined $\underset{\sim}{K} = \underset{\sim}{k}_i - \underset{\sim}{k}_s$, the scattering vector of the sample. Here $\underset{\sim}{r}_j$ specifies the position of the center of mass of the jth particle. By selecting the scattering angle θ_s, we have chosen a particular scattering vector for study. It is appropriate to view this experiment as Bragg scattering from the component of density fluctuation, which is propagating in the direction of $\underset{\sim}{K}$ and which has the wavelength $\Lambda_f = 2\pi/|\underset{\sim}{K}|$. The crests, which represent maximum particle density, are analogous to the scattering planes of atoms in X-ray crystallography. In order to obtain an expression for the magnitude of $\underset{\sim}{K}$, we note that in quasi-elastic light scattering the wavelength of the scattered light is almost unchanged, i.e., $|\underset{\sim}{k}_i| = |\underset{\sim}{k}_s| = 2\pi/\lambda$. The use of the construction in Figure 7 with elementary trigonometry yields the magnitude f $\underset{\sim}{K}$.

$$|\underset{\sim}{K}| = \frac{4\pi n}{\lambda_0} \sin(\theta_s/2) \tag{38'}$$

The angular dependence in light beating spectroscopy enters through this relation.

The total scattered field can now be written as

$$E_s = \sum_{j=1}^{N} A_j e^{i\underset{\sim}{K} \cdot \underset{\sim}{r}_j} E_0 e^{-i\omega_0 t} \tag{77}$$

where A_j contains the amplitude factors for the jth particle, which were discussed in Section II.A. In the notation of Section II.C, A_j is equal to aP for a molecule containing P segments. The intensity is given by

$$I_s = E_s E_s^* = \sum_{\ell,m=1}^{N} A_\ell A_m^* e^{i\underset{\sim}{K} \cdot (\underset{\sim}{r}_\ell - \underset{\sim}{r}_m)} E_0^2 \tag{78}$$

If the scattering particles are in motion, each $\underset{\sim}{r}_j$ is time dependent; and I_s fluctuates in time. From Equation 78 the time average of $I_s(t)$ is

$$\langle I_s(t) \rangle = \left[\langle \sum_{j=1}^{N} |A_j|^2 \rangle + \sum_{\ell \neq m}^{N} A_\ell A_m^* \langle e^{i\underset{\sim}{K} \cdot (\underset{\sim}{r}_\ell - \underset{\sim}{r}_m)} \rangle \right] E_0^2 \tag{79}$$

For particles that are identical, but independent, the second term on the right-hand side (rhs) of Equation 79 vanishes since the exponent is randomly distributed; and the equation becomes

$$<I_s(t)> = N<|A_j|^2> E_0^2 \tag{80}$$

3. Correlation Functions

The scattered field E_s is a random variable because of the time dependence implicit in Equation 77. For example, the position vectors $\underset{\sim}{r}_j$ depend on time. Also, if the scatterers are not isotropic, A_j depends on the orientation; and rotational motions produce fluctuations in E_s. Finally, the number of particles N in the scattering volume fluctuates about its mean value <N>. The latter effect is only important at very low concentrations and can usually be neglected for molecular solutes. It should be noted that N refers to the number of particles in the small volume defined by the laser beam and the imaging optics, while in Section II, N referred to the number of particles per unit volume. The most useful way to characterize the time dependence of $E_s(t)$ is through the first-order field autocorrelation function $G^{(1)}(\tau)$ defined by

$$G^{(1)}(\tau) = <E_s^*(t)E_s(t+\tau)> = <E_s^*(O)E_s(\tau)> \tag{81}$$

The rhs. of Equation 81 represents the time average of the product of the complex conjugate of the field at one time with the field at a time τ later. It is permissible to replace t in Equation 81 with zero, since $E_s(t)$ is a stationary random variable, and its properties are independent of the time origin. For $\tau = 0$, $G^{(1)}(\tau) = <I_s>$, while for large values of τ, $E_s(t)$ and $E_s(t + \tau)$ are uncorrelated and $G^{(1)}(\tau)$ approaches $<E_s(t)>^{|2}$, which is equal to zero.

For the scattered field E_s described in Equation 77 with $N = <N>$, Equation 81 gives:

$$G^{(1)}(\tau) = \sum_{\ell,m=1}^{N} <A_\ell^*(O)A_m(\tau)e^{i\underset{\sim}{K}\cdot[\underset{\sim}{r}_m(\tau) - \underset{\sim}{r}_\ell(O)]}> E_o^2 e^{-i\omega_0\tau} \tag{82}$$

If the motions of the different particles are uncorrelated and the translational and rotational motions are independent, Equation 82 reduces to:

$$G^{(1)}(\tau) = N<A_\ell^*(O)A_\ell(\tau)><e^{i\underset{\sim}{K}\cdot[\underset{\sim}{r}_\ell(\tau) - \underset{\sim}{r}_\ell(O)]}> E_o^2 e^{-i\omega_0\tau} \tag{83}$$

The summation has been dropped and the factor N inserted, since each particle has the same statistical behavior. The subscript ℓ is superfluous and can also be dropped. Thus, the field correlation function $G^{(1)}(\tau)$ is found to be proportional to the product of single particle correlation functions for rotation and translation.

The scattered intensity $I_s(t)$ is usually detected with a photomultiplier tube (PMT), and fluctuations in $I_s(t)$ appear either as fluctuations in the photocurrent i(t) or as fluctuations in the photocount rate. Thus with an ideal detection system

$$I_s(t) \propto i(t) \propto n(t,\Delta T)$$

Here $n(t,\Delta T)$ means the number of photons detected in the time interval from t to t + ΔT. The measured quantity is always $I_s(t)$ rather than $E_s(t)$, and the directly measured autocorrelation function is the second order function $G^{(2)}(\tau)$ defined by

$$G^{(2)}(\tau) = <I_s(O)I_s(\tau)> \tag{84}$$

In general $G^{(2)}(\tau)$ is not simply related to $G^{(1)}(\tau)$; however, in special cases useful relations exist. For example, if $E_s(t)$ is a Gaussian random variable the Siegert relation gives

$$G^{(2)}(\tau) = <I_s>^2 + |G^{(1)}(\tau)|^2 \tag{85}$$

Practically speaking this equation holds for scattering from solutions at room temperature, except at such low concentrations that number fluctuations become important (see Section III.F). The derivation of Equation 85 is not trivial, but elementary discussions are available.[35] It is conventional to define the reduced first and second order correlation functions $g^{(1)}(\tau)$ and $g^{(2)}(\tau)$, respectively, through the equations

$$g^{(1)}(\tau) = <E_s^*(O)E_s(\tau)>/<I> \tag{86}$$

$$g^{(2)}(\tau) = <E_s^*(O)E_s(O)E_s^*(\tau)E_s(\tau)>/<I>^2 \tag{87}$$

To see how an approximation to $G^{(2)}(\tau)$ can be derived from experimental data, consider Figure 8. The intensity is measured at time intervals Δt, and $G^{(2)}(\tau)$ is calculated through the relation

$$G^{(2)}(\tau) = \lim_{n\to\infty} \frac{1}{n} \sum_{i=O}^{n-1} I_s(t_i)I_s(t_{i+j}) \tag{88}$$

where $\tau = t_{i+j} - t_i = j\Delta t$. Commercial digital autocorrelators are available, which simultaneously calculate the products for all required values of τ and average these products to obtain an accurate representation of $G^{(2)}(\tau)$ in a few seconds if the intensity is sufficiently large.

While the function $G^{(2)}(\tau)$ is not familiar to most biochemists, its major features are easy to grasp. First consider the limits of short and long times.

$$\lim_{\tau\to O} <I_s(O)I_s(\tau)> = <I_s^2(O)> \tag{89}$$

$$\lim_{\tau\to\infty} <I_s(O)I_s(\tau)> = <I_s(O)>^2 \tag{90}$$

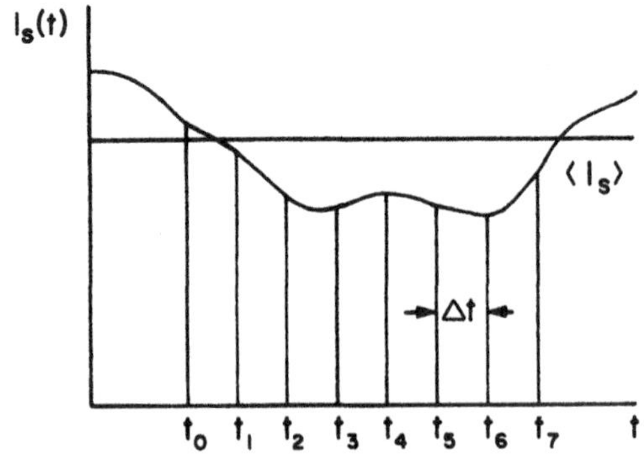

FIGURE 8. The scattered intensity $I_s(t)$ vs. time.

At the short time limit we obtain simply the average of $I^2_s(O)$. This is the largest value that $G^{(2)}(\tau)$ can attain. As τ increases the correlation between $I_s(t)$ and $I_s(t+\tau)$ decreases until the two values are completely independent. The average of their product becomes equal to the product of their averages. This behavior is illustrated in Figure 9 which shows a typical plot of $G^{(2)}(\tau)$ vs. τ. The monotonically decreasing function of τ is, for example, encountered for scattering particles that are undergoing Brownian motion. This particular case is discussed in detail in Section III.B. In other types of experiments oscillations are sometimes found in $G^{(2)}(\tau)$, but always the condition

$$G^{(2)}(\tau) \overline{<} <I_s^2(O)> \tag{91}$$

is satisfied.

4. The Frequency Spectrum

The rate at which $G^{(2)}(\tau)$ decreases indicates how rapidly $I_s(t)$ fluctuates. Rapid fluctuations in turn indicate that high frequency components are present, i.e., a rapid decay in the time domain corresponds to a broad line in the frequency domain. The quantitative expression of these ideas is given in Appendix E, which relates the power spectrum of I_s to its correlation function.

$$I_s^{(2)}(\omega) = \frac{\text{Re}}{\pi}\int_0^{\infty} G^{(2)}(\tau)e^{i\omega\tau}\,d\tau \tag{92}$$

In experimental situations, $I_s^{(2)}(\omega)$ can be obtained directly by analysis of the PMT photocurrent using a spectrum analyzer. It should be realized that $I_s^{(2)}(\omega)$ is not equivalent to the usual optical spectrum $I_s^{(1)}(\omega)$, which is obtained by filter type optical

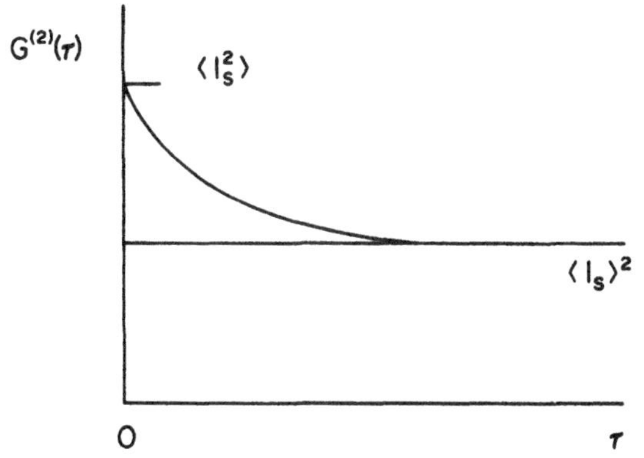

FIGURE 9. Typical behavior of the second-order correlation function $G^{(2)}(\tau)$ vs. τ.

instruments such as grating monochrometers. However, $I_s^{(1)}(\omega)$ is related to $G^{(1)}(\tau)$ and can in principle be calculated using the results of Appendix E.

$$I_s^{(1)}(\omega) = \frac{\mathrm{Re}}{\pi} \int_0^{\infty} G^{(1)}(\tau) e^{i\omega\tau} d\tau \tag{93}$$

A very important feature of $G^{(2)}(\tau)$ is that E_s appears only in products with its complex conjugate. Factors involving the angular frequency ω_o of the incident light cancel, and only fluctuations in the audio range remain. Accordingly, the spectrum $I^{(2)}(\omega)$ is referred to zero frequency. In contrast to this $G^{(1)}(\tau)$ still contains ω_o, and the frequencies in $I_s^{(1)}(\omega)$ are referenced to ω_o.

B. Translational Diffusion

1. The Diffusion Equation

Probably the most important application of photon correlation spectroscopy is the study of translational diffusion for macromolecules in solution.[36] This technique often permits the z-average diffusion coefficient D_T to be measured quickly and accurately. D_T directly gives information about molecular diameters and when combined with other results, e.g., the Svedberg constant, permits the molecular weights to be determined. The equations that relate D_T to the measured intensity correlation function are based on Equation 83 for $G^{(1)}(\tau)$. For small isotropic scatters Equations 80 and 83 can be combined to obtain the reduced correlation function $g^{(1)}(\tau)$ as defined in Equation 86.

$$g^{(1)}(\tau) = \langle e^{i\underset{\sim}{K} \cdot [\underset{\sim}{r}_\ell(\tau) - \underset{\sim}{r}_\ell(0)]} \rangle e^{-i\omega_o\tau} \tag{94}$$

The heart of the problem is the evaluation of the function $F_s(K,t) = \langle e^{i\underset{\sim}{K}\cdot\underset{\sim}{r}}\rangle$ where for simplicity we have introduced $r = \underset{\sim}{r}_\ell(t) - \underset{\sim}{r}_\ell(O)$. For translational diffusion the appropriate average can be calculated using the weighting factor $P(0|\underset{\sim}{r},t)$, which is the conditional probability that a particle will be found in the volume element d^3r at the position $\underset{\sim}{r}$ at time t if it were located at $\underset{\sim}{r} = 0$ initially. Thus

$$F_s(\underset{\sim}{K},t) = \int_0^\infty P(0|\underset{\sim}{r},t)e^{i\underset{\sim}{K}\cdot\underset{\sim}{r}}d^3r \tag{95}$$

It should be recognized that $F_s(\underset{\sim}{K},t)$ is just the spatial Fourier transform of $P(0|\underset{\sim}{r},t)$.

According to Fick's first law of diffusion, the particle flux $\underset{\sim}{J}$, i.e., the rate of flow of mass at $\underset{\sim}{r}$, is proportional to the gradient of the concentration.

$$\underset{\sim}{J}(\underset{\sim}{r},t) = -D_T\underset{\sim}{\nabla}C(\underset{\sim}{r},t) \tag{96}$$

The continuity equation, which assures the conservation of mass, is

$$\frac{\partial C(\underset{\sim}{r},t)}{\partial t} = -\underset{\sim}{\nabla}\cdot\underset{\sim}{J}(\underset{\sim}{r},t) \tag{97}$$

and this equation can be combined with the definition of $\underset{\sim}{J}$ to give Fick's second law of diffusion

$$\frac{\partial C(\underset{\sim}{r},t)}{\partial t} = D_T\nabla^2C(\underset{\sim}{r},t) \tag{98}$$

In this derivation it is assumed that the translational diffusion coefficient D_T is independent of concentration. This is not strictly true, and Equation 98 is only expected to be accurate in the limit of low concentrations. Equations for D_T and its concentration dependence are discussed in Appendix H. At low concentrations it is reasonable to assume that $P(0|\underset{\sim}{r},t)$ also obeys the diffusion equation so that

$$\frac{\partial P(0|\underset{\sim}{r},t)}{\partial t} = D_T\nabla^2P(O|\underset{\sim}{r},t) \tag{99}$$

The brute force procedure is to solve Equation 99 for $P(0|\underset{\sim}{r},t)$ with the initial condition $P(0|r,0) = \delta(\underset{\sim}{r})$ where $\delta(\underset{\sim}{r})$ is the Dirac delta function. $P(0|\underset{\sim}{r},t)$ is then substituted into Equation 95, and the integral is evaluated to obtain $F_s(\underset{\sim}{K},t)$. It is much simpler to use Fourier transform methods to obtain $F(K,t)$ directly from Equation 99. We first multiply both sides of Equation 99 by $e^{i\underset{\sim}{K}\cdot\underset{\sim}{r}}$ and then integrate over volume to obtain

$$\int_{0}^{\infty} e^{i\underset{\sim}{K}\cdot\underset{\sim}{r}} \frac{\partial P(0|\underset{\sim}{r},t)}{\partial t} d^3r = D_T \int_{0}^{\infty} e^{i\underset{\sim}{K}\cdot\underset{\sim}{r}} \nabla^2 P(O|\underset{\sim}{r},t) d^3r$$

$$\frac{\partial F_s(\underset{\sim}{K},t)}{\partial t} = -D_T K^2 F_s(\underset{\sim}{K},t) \tag{100}$$

On the left-hand side (lhs) we have simply factored $\partial/\partial t$ from the integral and used Equation 95 for $F_s(\underset{\sim}{K},t)$. The rhs of Equation 100 follows from a property of Fourier transforms, namely,

$$\int_{-\infty}^{+\infty} e^{iKy} \frac{\partial^n}{\partial y^n} P(y)dy = (-iK)^n \int_{-\infty}^{+\infty} e^{iKy} P(y)dy \tag{101}$$

which can be proved using integration by parts.

Equation 100 is a first order differential equation that can easily be integrated to give

$$F_S(\underset{\sim}{K},t) = \underset{\sim}{F}_S(\underset{\sim}{K},0)\, e^{-D_T K^2 t} \tag{102}$$

The appropriate initial condition is $F_s(\underset{\sim}{K},0) = 1$, which can be derived either by referring to the definition of $F_s(\underset{\sim}{K},t)$ in Equation 94 or by substituting the initial condition $P(0|\underset{\sim}{r},0) = \delta(\underset{\sim}{r})$ into Equation 95. Now having Equation 102 for $F_s(\underset{\sim}{K},t)$ we return to Equation 94 to obtain the desired expression

$$g^{(1)}(\tau) = \frac{G^{(1)}(\tau)}{<I_s>} = e^{-D_T K^2 \tau}\, e^{-i\omega_0 \tau} \tag{103}$$

The optical spectrum associated with $G^{(1)}(\tau)$ can be derived using Equation 93. Thus,

$$\begin{aligned} I^{(1)}(\omega) &= \frac{\text{Re}}{\pi} \int_{0}^{\infty} \left[<I_S> e^{-D_T K^2 \tau}\, e^{-i\omega_0 \tau}\right] e^{i\omega\tau} d\tau \\ &= \frac{<I_s>}{\pi} \left[\frac{D_T K^2}{(D_T K^2)^2 + (\omega_0 - \omega)^2}\right] \end{aligned} \tag{104}$$

This equation describes a Lorentzian curve centered at the laser frequency ω_0 and having a half-width at half-height (HWHH) of $D_T K^2$ (see Figure 10a). However, as previously pointed out, ω_0 is at least 10^{10} tmes larger than the linewidth if the scattering particles are macromolecules; and it is impossible with optical techniques to measure the linewidth. Experiments which do permit D_T to be determined are based on $G^{(2)}(\tau)$.

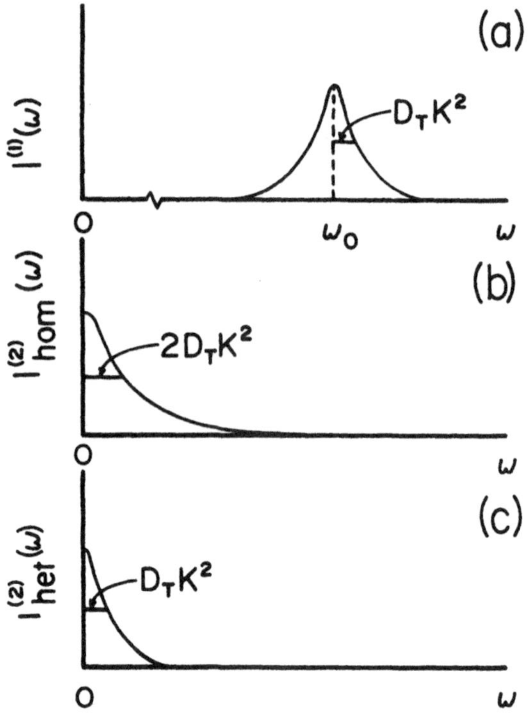

FIGURE 10. Calculated spectra for light scattered from diffusing particles in three cases: (a) the optical spectrum, (b) the power spectrum in a homodyne experiment, and (c) the power spectrum in a heterodyne experiment.

2. Homodyne Experiment

In this experiment the scattered light is detected with a PMT and the photon count rate or the photocurrent is used to generate an approximation to $G^{(2)}(\tau)$. The Siegert relation, Equation 85, is then used in the modified form:

$$\frac{G^{(2)}(\tau)}{\langle I_S \rangle^2} = g^{(2)}(\tau) = 1 + \beta |g^{(1)}(\tau)|^2 \qquad (105)$$

where β is a factor of order unity, which accounts for the temporal and spatial integrations which are unavoidable in real experiments.[37] Approximations to β can be derived, but it is usually just taken to be an instrumental parameter which is maximized, if possible, and then measured in each experiment. The output of an autocorrelator provides sufficient information for the calculation of $g^{(2)}(\tau)$. This reduced correlation function is then analyzed using Equation 105 in the form:

$$Y(\tau) = \ln \left[g^{(2)}(\tau) - 1 \right]^{1/2} = \frac{1}{2} \ln \beta + \ln |g^{(1)}(\tau)| \qquad (106)$$

For translational diffusion in a monodisperse solution Equation 103 is introduced for $g^{(1)}(\tau)$ to give

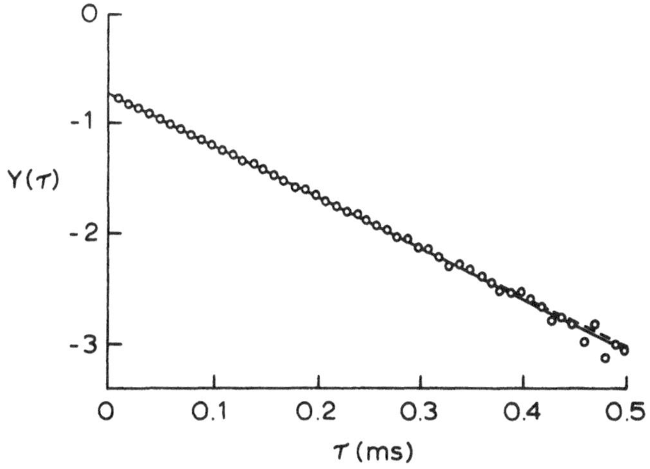

FIGURE 11. Experimental values of $Y(\tau)$ plotted vs. τ for a sample of oxyhemoglobin A at a concentration of 16.3 g/dl and T = 20°C. The curves represent the best linear (solid) and best quadratic (dashed) fits to the data (see Equations 107 and 115). (From Jones, C. R., Ph.D. thesis, University of North Carolina, Chapel Hill, 1977.)

$$Y(\tau) = \frac{1}{2}\,\ell n\,\beta - D_T K^2 \tau \qquad (107)$$

A plot of $Y(\tau)$ vs. τ then gives a straight line having the slope $-D_T K^2$ and the intercept $(\ell n\beta)/2$. An example of such a plot for oxy-hemoglobin A is shown in Figure 11.[38] The small deviation from linearity indicates polydispersity and suggests that Equation 107 should have additional terms on the rhs which depend on higher powers of τ.

The power spectrum associated with $G^{(2)}(\tau)$ can be obtained using Equations 92, 103, and 105.

$$\begin{aligned} I^{(2)}_{hom}(\omega) &= \frac{Re}{\pi}\int_0^{\infty} \langle I_s\rangle^2 \left[1 + \beta e^{-2D_T K^2 \tau}\right] e^{i\omega\tau}\, d\tau \\ &= \langle I_S\rangle^2 \delta(\omega) + \frac{\beta\langle I_S\rangle^2}{\pi}\left[\frac{2D_T K^2}{(2D_T K^2)^2 + \omega^2}\right] \end{aligned} \qquad (108)$$

This equation represents a spike (delta function) at ω = O and a superimposed Lorentzian curve having a HWHH of $2D_T k^2$ centered at ω = O (see Figure 10B). The power spectrum lies in the audio range and can be obtained directly from the fluctuating photocurrent by means of an electronic spectrum analyzer. In fact, light-beating experiments were first done this way before correlators were available. An example of an experimental power spectrum is shown in Figure 12 for a 1% solution of RNase A at θ_S = 90° and T = 24°C.[39] This experiment in principle gives the same information as that obtained from $G^{(2)}(\tau)$; however, in practice the power spectrum sometimes gives less warning of polydispersity. For example the data points in Figure 12 fit the calculated Lorentzian curve within experimental error even though the sample contains dimers and tetramers as well as monomers. The measured linewidth (HWHH) for this sample is 7.63 kHz, while that obtained for a solution containing only monomers is

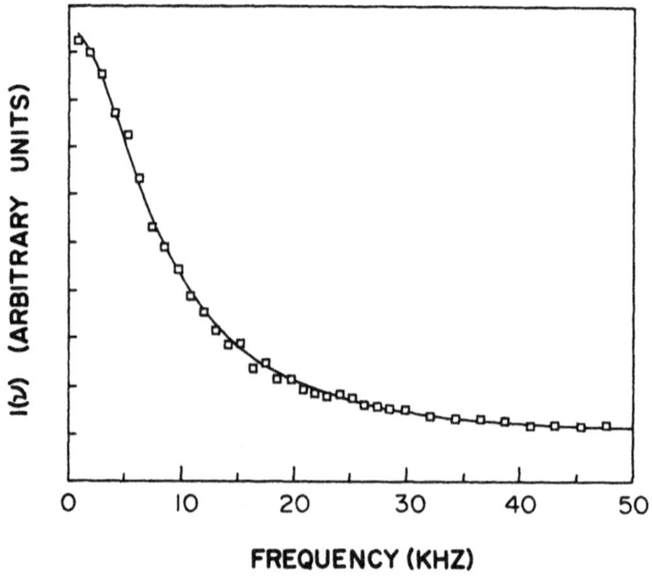

FIGURE 12. The power spectrum $I^{(2)}(\omega)$ of a 1% solution of RNase A measured at $\theta_s = 90°$ and T = 24°C. The solid curve is the best single Lorentzian fit to the experimental data (every fourth data point is displayed). The HWHH value is 7.63 KHz. (From Wilson, W. W., Ph.D. thesis, University of North Carolina, Chapel Hill, 1973.)

12.9 kHz corresponding to $D_T = 11.7 \times 10^{-7}$ cm²/sec at 20°C. In this case the determination of the fraction of monomers required the use of chromatography.

3. Heterodyne Experiment

In this experiment a reference beam is obtained either by deflecting a small fraction of the intensity of the laser light before it reaches the sample or by scattering part of the incident light from a solid object in the scattering volume, e.g., a needle. In either case the reference beam is mixed with the light scattered from the sample on the PMT. The reference beam is called the local oscillator, and its electric field and intensity are denoted by $E_L(t)$ and $I_L(t)$, respectively. The total electric field $E_{det}(t)$ of the light at the PMT detector is the sum of contributions from the local oscillator and light scattered from the sample. Thus

$$E_{det}(t) = E_S(t) + E_L(t) \tag{109}$$

and the associated intensity correlation function is

$$G^{(2)}(\tau) = \langle E^*_{det}(t)\, E_{det}(t)\, E^*_{det}(t+\tau)\, E_{det}(t+\tau)\rangle \tag{110}$$

This function is considered in Appendix F for the special case that $I_L >> \langle I_S \rangle$, and it is shown that

$$G^{(2)}(\tau) = I_L^2 + 2\,I_L\,\mathrm{Re}\left[G_S^{(1)}(\tau)\,e^{i\omega_0\tau}\right] \tag{111}$$

where $G_s^{(1)}(\tau) = \langle E^*_s(O)\,E_s(\tau)\rangle$. When the intensity fluctuations result from translational diffusion, $G_s^{(1)}(\tau)$ is given by Equation 103, and 111 becomes

$$G^{(2)}(\tau) = I_L^2 + 2\,I_L\,\langle I_S\rangle\,e^{-D_T K^2 \tau} \tag{112}$$

In contrast to the $G^{(2)}(\tau)$ function which was derived for the homodyne experiment, here the background (base line) depends on the intensity of the reference beam, and the time dependent term contains $\mathrm{Re}[G_s^{(1)}(\tau)]$ rather than $|G_s^{(1)}(\tau)|^2$.

The power spectrum associated with Equation 112 is obtained by using Equation 92,

$$\begin{aligned} I_{het}^{(2)}(\tau) &= \frac{\mathrm{Re}}{\pi}\int_0^\infty \left[I_L^2 + 2\,I_L\,\langle I_S\rangle\,\mathrm{Re}\left(e^{-D_T K^2 \tau}\right)\right] e^{i\omega\tau}\,d\tau \\ &= I_L^2\,\delta(\omega) + \frac{2}{\pi}\,I_L\,\langle I_S\rangle\,\mathrm{Re}\int_0^\infty e^{-D_T K^2\tau}\,e^{i\omega\tau}\,d\tau \\ &= I_L^2\,\delta(\omega) + \frac{2}{\pi}\,I_L\,\langle I_S\rangle\left[\frac{D_T K^2}{(D_T K^2)^2 + \omega^2}\right] \end{aligned} \tag{113}$$

The result is a delta function at zero frequency and a Lorentzian curve having the same width as the optical spectrum but also centered at zero frequency rather than at the frequency of the incident light (see Figure 10c). The linewidth is thus a factor of two, smaller in the heterodyne experiment than in the homodyne experiment. There is usually no particular advantage to using the heterodyne arrangement when translational diffusion is being studied, but there is the danger that an unknown amount of the heterodyne component may be present in what is thought to be a homodyne experiment because of stray light or because of dust particles in solution. However, when the scattering particles are undergoing uniform motion, e.g., in flowing liquids or in electrophoresis experiments, the Doppler effect associated with the velocity of the particles relative to the source and the detector can be measured in the heterodyne experiment, but not in the homodyne experiment. This forms the basis of laser velocimetry, some applications of which are discussed in Section III.C.

4. Data Analysis and Experimental Results

Quasi-elastic scattering has become the standard method for the measurement of diffusion coefficients for macromolecules. Any of the arrangements discussed above can be used to measure D_T; however, the determination of $G^{(2)}(\tau)$ by means of the homodyne experiment is the method of choice. Major reasons for this are experimental convenience and ease of data analysis. Since all samples are polydisperse to some extent, the function $|g^{(1)}(\tau)|$, required in Equation 106, must in general be written as

$$|g^{(1)}(\tau)| = \int_0^\infty G(\Gamma)\, e^{-\Gamma\tau} d\Gamma \tag{114}$$

where $G(\Gamma)$ is the normalized distribution function for the decay constants and $\Gamma = D_T K^2$. The consequence of the distribution is that Equation 107 must be replaced with the cumulant expansion[40]

$$Y(\tau) = \frac{1}{2} \ln \beta + \sum_{m=1}^{\infty} K_m \frac{(-\tau)^m}{m!} \tag{115}$$

It is shown in Appendix G that the first four cumulants K_m are

$$K_1 = <\Gamma>, K_2 = \mu_2, K_3 = \mu_3, K_4 = \mu_4 - 3(\mu_2)^2 \tag{116}$$

where the mth moment of the distribution is defined by

$$\mu_m = \int_0^\infty (\Gamma - <\Gamma>)^m G(\Gamma) d\Gamma \tag{117}$$

and

$$<\Gamma> = \int_0^\infty \Gamma G(\Gamma) d\Gamma \tag{118}$$

The z-average diffusion coefficient is given by $<D_T> = <\Gamma>/K^2$, the standard deviation of the distribution is specified by $\sqrt{\mu_2}$, the skewness by $\mu_3/\mu_2^{3/2}$, etc. The normalized variance $\mu_2/<\Gamma>^2$ is often quoted as a measure of polydispersity.

Several schemes have been proposed for determining the moments from experimental data. For example, since

$$(-1)^n \frac{d^n}{d\tau^n} Y(\tau) \Big|_{\tau = 0} = K_n \tag{119}$$

a plot of the second derivative of $Y(\tau)$ with respect to τ vs. τ can be extrapolated to $\tau = 0$ to obtain μ_2. For samples which are supposed to be monodisperse it is also reasonable to fit $Y(\tau)$ to a quadratic function in τ. According to Equation 115 the linear term contains $<D_T>$ and the coefficient of τ^2 can be used to calculate the variance. If the variance is greater than 0.02, an artifact such as dust in the sample or stray light is probably present. In principle the variance should be useful in the determination of equilibrium constants for dissociation reactions. However, a simple calculation for monomer-dimer equilibria shows that the variance never exceeds ~0.014.

Experience shows that $\langle D_T \rangle$ can be measured in a few minutes with an error of only about 1% for nonabsorbing samples. Variances greater than 0.1 can be measured with fair accuracy, but the higher cumulants have very large uncertainties. Further progress can be made only if the functional form of $G(\Gamma)$ is known or assumed.[41] Additional information about polydispersity can be obtained from a plot of $\langle \Gamma \rangle$ vs. K^2; therefore, a determination of the angular dependence of $\langle \Gamma \rangle$ is always recommended.

The translational diffusion coefficient D_T was introduced in Equation 96. For a sphere in a viscous medium, D_T is given by

$$D_T = \frac{k_B T}{f_T} = \frac{k_B T}{6\pi\eta a_h} \tag{120}$$

where f_T is the translational friction coefficient, η is the coefficient of viscosity of the solvent, and a_h is the hydrodynamic radius of the sphere. The derivation of this equation, which was first presented by Einstein,[42] is discussed in Appendix H. Equation 120 permits the calculation of molecular radii; however, it should be kept in mind that the hydrodynamic radius may include a hydration layer. Also, it is clear from experimental results that D_T depends on the average concentration $C(\underset{\sim}{r},t) \equiv c$. The concentration dependence enters the theory in two ways. First, the derivation of Equation 120 requires that the gradient of the chemical potential μ_2, which is the driving force of the diffusion, be related to the gradient of the concentration of the solute. At finite concentrations this introduces a virial expansion in the numerator of Equation 120. And second, the friction coefficient f depends on the concentration and it can be expanded in terms of the concentration. The result is that D_T can be written as (see Appendix H)[43]

$$D_T(C) = \frac{k_B T}{f_0} \frac{(1 + 2B_2\, MC + 3B_3\, MC^2 + \ldots)}{(1 + B'\, C + \ldots)} \tag{121}$$

where the B_i are the virial coefficients which occur in the expansion of the chemical potential μ_1 of the solvent in terms of C. For small concentrations Equation 121 has the form

$$D_T(C) = D_0\,(1 + K_D\, C + \ldots) \tag{122}$$

where $K_D = (2MB_2 - B')$. The partial cancellation which occurs here probably accounts for the small concentration dependence which is often found for diffusion coefficients. Numerous theoretical treatments have been presented which yield equations for D_T similar to Equation 122, but with different values of K_D. There is still no consensus on the proper way to include hydrodynamic interactions in the calculation of K_D.

Table 3
TRANSLATIONAL DIFFUSION COEFFICIENTS FROM QUASI-ELASTIC LIGHT SCATTERING FOR SELECTED SYSTEMS

Macromolecule	Conditions	$D_T(10^{-7}\ cm^2/s)$	Ref.
Myosin (monomer)		1.24	46
(dimer)		0.84	46
Fibrinogen		2.04 ± 0.09	47
γ-Globulin		3.8	43
Bovine glutamate dehydrogenase (GDG) monomer, pH 7		4.51[a]	48
Oxy-hemoglobin (oxy-Hb)	pH 6.9 (HbA); pH 7.1 (HbS); 0.1 *M* KCl	6.9[b]	49
Bovine serum albumin (BSA)	pH 6.91; no salt	10.2[c]	50
	pH 6.8; 0.5 *M* KCl	6.7	50
Lysozyme	pH 4.2; sodium acetate-acetic acid buffer	10.6[d]	51
Ribonuclease A (RNase)	pH 8.1; 0.5 *M* KCl; 24°C	12.6	52
Avian myeloblastosis virus (AMV)		0.268	53
Vesicular stomatitis virus (VSV)		0.29	54
Rous sarcoma virus (RSV)		0.291	53
Tobacco mosaic virus (TMV)		0.39	55
Infectious pancreatic nucrosis virus (IPNV)		0.67	56
Turnip yellow mosaic virus (TYMV)		1.44	56
R17		1.534 ± 0.015	57
Qβ		1.423 ± 0.014	57
BSV		1.246 ± 0.013	57
PM2		0.650 ± 0.007	57
T2		0.644 ± 0.007	57
Ribosomes 70S		1.81	58
50S		2.1	58
30S		2.2	58

[a] Corrected to water at 25°C.
[b] Corrected to water at 20°C.
[c] Corrected to water at 25°C.
[d] Corrected to water at 20°C.

We emphasize that quasi-elastic light scattering detects mutual diffusion, which is driven by a concentration gradient of solute molecules. This is to be contrasted with tracer diffusion where the solute concentration is uniform, but the migration of "labeled" solute molecules can be followed. The classical diaphragm method in which diffusion through a millipore membrane is measured, while satisfactory for the measurement of tracer diffusion coefficients, is apparently invalid for mutual diffusion measurements on protein solutions because of the osmotic back flow of solvent across the membrane.[44,116] This explains the marked discrepancies between the concentration dependences of D_T obtained by quasi-elastic light scattering and by the membrane method.[45,49] For charged macromolecules the situation is even more complicated. One important consequence of long range interactions is a $\underline{K}$ dependence of the apparent diffusion coefficient. Recent work in this area is reviewed in Reference 7.

The number of diffusion coefficients determined by quasi-elastic light scattering is now very large, and we only list a few representative examples in Table 3. In reporting diffusion coefficients, it is common to correct the measured values of D_T so that they represent the hypothetical case of the same hydrodynamic particle moving in pure water at 20°C. This is accomplished using Equation 120 and the fact that $f = \eta(T)G$

where $\eta(T_1)$ is the viscosity of the solvent and G is a form factor, which is characteristic of the molecule.[59] For macromolecules in salt solutions this treatment assumes that there are only two components, i.e., the solvent is taken to be the salt solution without the macromolecule and $\eta(T)$ refers to the viscosity of this salt solution. Thus, if D_T is the measured quantity at the temperature T, the correction gives

$$D_T(20°C, \text{water}) = D_T(T, \text{salt sol}) \left[\frac{293.2}{T(°K)}\right] \left[\frac{\eta(T, \text{salt sol})}{\eta(20°C, \text{water})}\right] \tag{123}$$

where $\eta(20°, \text{water}) = 1.008$ cP.

It should be noted that accurate diffusion constants can be obtained for large particles such as viruses and ribosomes. This is important in the determination of weights of such particles since their diffusion coefficients are not easily obtained by other methods. The procedure is to combine the measured z-average diffusion coefficient D_T with the sedimentation coefficient S_o in the Svedberg equation to obtain the weight average molecular weight in the limit of infinite dilution.[60]

$$M_w = \frac{RT\, S_o}{D_T(1 - \bar{v}\rho)} \tag{124}$$

Here $\bar{v}$ is the partial specific volume of the solute and ρ is the density of the solvent. For example with reovirus, where $S_o = 734 \times 10^{-13}$, $D_T = 0.44 \times 10^{-7}$ cm²/s, and $\bar{v} = 0.690$ mℓ/g, Equation 124 gives $M_w = 12.5 \times 10^6$.[56]

C. Directed Flow

1. Constant Velocities and Laser Velocimetry

Frequency shifts for light scattered from particles in uniform motion are conceptually simpler than for diffusing particles. The reason for this is that the concept of the Doppler shift can be applied more directly for constant velocity. Consider for example the situation in Figure 13. The angular frequency of the light emitted from the laser is ω_o; however, the apparent frequency seen by the scattering particle is Doppler shifted by the amount $\Delta\omega = -(n\underset{\sim}{v}_d \cdot \hat{e}/c)\omega_o$ where $\hat{e}$ is a unit vector in the direction of $\underset{\sim}{k}_i$ and $\underset{\sim}{v}_d$ is the velocity of the scattering particle. A second Doppler shift enters the measurement since the scattering particle is also moving relative to the detector. The situation can be summarized as follows using $n\omega_o/c = |\underset{\sim}{k}_i| = |\underset{\sim}{k}_s|$.

1. Source frequency: ω_o
2. Frequency seen by moving particle $\omega_o - \underset{\sim}{k}_i \cdot \underset{\sim}{v}_d$
3. Frequency seen by detector $(\omega_o - \underset{\sim}{k}_i \cdot \underset{\sim}{v}_d) + \underset{\sim}{k}_s \cdot \underset{\sim}{v}_d$

Thus the detected frequency is shifted relative to the source frequency by the amount $\Delta\omega = -\underset{\sim}{K} \cdot \underset{\sim}{v}_d$. This result is, in fact, already contained in Equation 83 for $G^{(1)}(\tau)$ since for uniform motion $\underset{\sim}{r}_\ell(\tau) - \underset{\sim}{r}_\ell(O) = \underset{\sim}{v}_d\tau$. The correlation function is then

$$G^{(1)}(\tau) = N \langle |A_\ell|^2 \rangle e^{i\underset{\sim}{K} \cdot \underset{\sim}{v}_d \tau} E_o^2 \, e^{-i\omega_o \tau} \tag{125}$$

which is equivalent to replacing the source frequency of $\omega_o - \underset{\sim}{K} \cdot \underset{\sim}{v}_d$.

As previously discussed, the magnitude of $\underset{\sim}{v}_d$ expected for a macromolecule corresponds to such a small frequency shift that a light-beating experiment must be used to measure $\Delta\omega$. Suppose that a homodyne experiment is employed. In this case Equations 85 and 125 combine to show that $G^{(2)}_{hom}(\tau) = 2 <I_s>^2$, i.e., the intensity correlation function contains no information at all about $\underset{\sim}{v}_d$! The reason for this is that in a homodyne experiment the phase of the scattered wave does not affect the measured intensity. In order to observe the effect of $\underset{\sim}{v}_d$, a fixed reference frequency must be introduced, so that interference terms will alter the intensity, and this is just what is done in the heterodyne experiment. The information content of the heterodyne experiment can be seen by combining Equation 125 for $G^{(1)}(\tau)$ with Equation 111 to obtain

$$\begin{aligned} G^{(2)}_{het}(\tau) &= I_L^2 + 2 I_L \,\mathrm{Re}[N<|A_{\underset{\sim}{Q}}|^2> e^{i\underset{\sim}{K}\cdot\underset{\sim}{v}_d\tau}]\, E_0^2 \\ &= I_L^2 + 2 I_L <I_S> \cos(\underset{\sim}{K}\cdot\underset{\sim}{v}_d\tau) \end{aligned} \qquad (126)$$

The oscillatory behavior exhibited by Equation 126 indicates that a frequency shift is present. The exact form of the frequency spectrum is obtained by substituting Equation 126 into Equation 92,

$$\begin{aligned} I^{(2)}_{het}(\omega) &= I_L^2\, \delta(\omega) + \frac{2 I_L}{\pi} <I_S> \mathrm{Re}\int_0^\infty \cos(\underset{\sim}{K}\cdot\underset{\sim}{v}_d\tau)\, e^{-i\omega\tau} d\tau \\ &= I_L^2\, \delta(\omega) + I_L <I_S> [\delta(\omega - \underset{\sim}{K}\cdot\underset{\sim}{v}_d) + \delta(\omega + \underset{\sim}{K}\cdot\underset{\sim}{v}_d)] \end{aligned} \qquad (127)$$

where, as in Equation 113, we have used the fact that the Fourier transform of unity is the delta function. Equation 127 shows that an infinitely narrow peak appears with a shift of $\Delta\omega = \underset{\sim}{K}\cdot\underset{\sim}{v}_d$ for either the velocity $\underset{\sim}{v}_d$ or $-\underset{\sim}{v}_d$. Perfectly uniform velocities are, of course, unattainable; and the distribution of velocities of the scattering particles will determine the width and shape of the shifted peak.

For application to real systems Equation 127 can be written as

$$I_{exp}(\omega) = \int P(v)\, I^{(2)}_{het}(\omega)\, dv \qquad (128)$$

The experimental lineshape $I_{exp}(\omega)$ can then be used to determine the velocity distribution function P(v). Two important applications of this equation have been to blood flow in vessels and protoplasmic streaming in living cells. For example, Tanaka et al.[61] measured the flow rate in retinal vessels by focusing a low power laser into the eye and analyzing the back scattered light. Also, Tanaka and Benedek[62] measured blood flow in the femoral vein of a rabbit by inserting a fiber optic catheter with a beveled and polished end. Laser light passed through the fiber into the plasma, and the light scattered by moving particles (erythrocytes) was collected by the same fiber. In the first of a series of investigations Mustacich and Ware[63] scattered light from the protoplasm in a *Nitella* cell and used the frequency spectrum to map the flow pattern and to determine the velocity distribution function.

A final comment concerning the inability of the homodyne experiment to detect the uniform velocity $\underset{\sim}{v}_d$ is in order. The homodyne experiment does contain information about diffusion since the ensemble average of the phase factor $e^{i\underset{\sim}{K}\cdot\underset{\sim}{r}}$ produces the real function $F_s(\underset{\sim}{K},t)$ as shown in Equation 102. In a monodisperse solution each scattering particle exhibits the same average behavior, and each term on the rhs of Equation 82 makes the same contribution. A quite different result is expected for scattering from a gas at low pressure. The jth molecule has the velocity $\underset{\sim}{v}_j$ and contributes the term

$$<|A_j|^2> e^{i\underset{\sim}{K}\cdot\underset{\sim}{v}_j\tau} E_0^2 e^{-i\omega_0\tau}$$

to the sum in Equation 82. The contributions from the different molecules are not the same; however, a Maxwell-Boltzmann distribution of velocities is present and the summation in Equation 82 can be performed without difficulty (see Section III.E). The weighted sum of the phase factors $e^{i\underset{\sim}{K}\cdot\underset{\sim}{v}_j\tau}$ produces a real function which contains the mean square velocity $<v^2>$. The homodyne experiment, therefore, permits the measurement of $<v^2>$ even though no information would have been obtained if all of the molecules had the same velocity vector $\underset{\sim}{v}$. The conclusion is that the wave scattered from particles having the velocity $\underset{\sim}{v}_i$ provides a reference for the wave scattered from particles having the velocity $\underset{\sim}{v}_j$. In fact, if only two velocities were present, the sum in Equation 82 would contain two terms and $|G^{(1)}(\tau)|^2$ would depend on the velocity difference $\underset{\sim}{v}_i - \underset{\sim}{v}_j$. This information, i.e., the quantity $<v^2>$, is obtained about the velocities only because more than one velocity is present. However, the average velocity $<v>$ is still inaccessible in the standard homodyne experiment.

2. Forced Diffusion

In a flowing solution of macromolecules or in electrophoresis where an electric field is applied, molecules undergo both directed flow and diffusion. The equation for forced diffusion can be used to describe this situation. The particle flux $\underset{\sim}{J}(\underset{\sim}{r},t)$ now contains the flow term $\underset{\sim}{v}_d C(\underset{\sim}{r},t)$ in addition to the diffusive term found in Equation 96. Thus

$$\underset{\sim}{J}(\underset{\sim}{r},t) = \underset{\sim}{v}_d\, C(\underset{\sim}{r},t) - D_T\, \underset{\sim}{\nabla}\, C(\underset{\sim}{r},t) \qquad (129)$$

This expression for $\underset{\sim}{J}(\underset{\sim}{r},t)$ can be substituted into Equation 97 to obtain:

$$\frac{\partial C(\underset{\sim}{r},t)}{\partial t} + \underset{\sim}{\nabla}\cdot[\underset{\sim}{v}_d\, C(\underset{\sim}{r},t) - D_T\, \underset{\sim}{\nabla}\, C(\underset{\sim}{r},t)] = 0 \qquad (130)$$

As in Section III.B, the conditional probability function $P(0|\underset{\sim}{r},t)$ is required so that the function $F_s(\underset{\sim}{K},t)$, which appears in $g^{(1)}(\tau)$, can be calculated. For dilute solutions $P(0|\underset{\sim}{r},t)$ obeys Equation 130, and we write

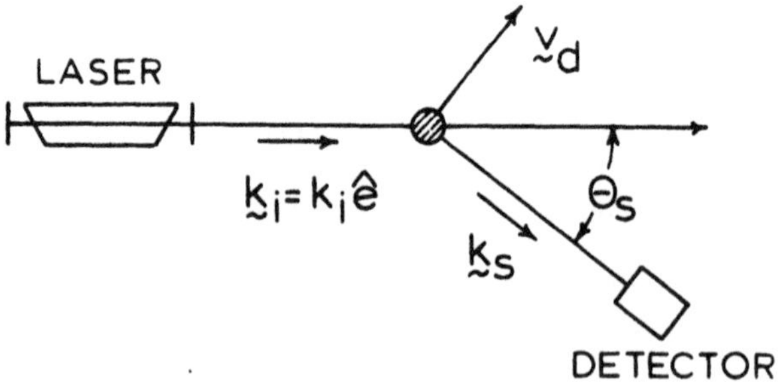

FIGURE 13. The Doppler scattering experiment for a particle having the constant velocity $\underset{\sim}{v}_d$.

$$\frac{\partial P(0|\underset{\sim}{r},t)}{\partial t} + \underset{\sim}{v}_d \cdot \underset{\sim}{\nabla} P(0|\underset{\sim}{r},t) - D_T \nabla^2 P(O|\underset{\sim}{r},t) = 0 \tag{131}$$

where the identity $\underset{\sim}{\nabla} \cdot \underset{\sim}{v}_d C = \underset{\sim}{v}_d \cdot \underset{\sim}{\nabla} C$ has been used. In solving for $F_s(\underset{\sim}{K},t)$ the same procedure is followed as in Equation 100. After multiplying Equation 131 by $e^{i\underset{\sim}{K}\cdot\underset{\sim}{r}}$ and integrating, we obtain

$$\frac{\partial F_S(\underset{\sim}{K},t)}{\partial t} + \underset{\sim}{v}_d \cdot \int_0^\infty e^{i\underset{\sim}{K}\cdot\underset{\sim}{r}} \underset{\sim}{\nabla} P(0|\underset{\sim}{r},t) d^3r + D_T K^2 F_s(\underset{\sim}{K},t) = 0 \tag{132}$$

and using Equation 101 this becomes

$$\frac{\partial F_S(\underset{\sim}{K},t)}{\partial t} - i(\underset{\sim}{v}_d \cdot \underset{\sim}{K}) F_s(\underset{\sim}{K},t) = -D_T K^2 F_S(\underset{\sim}{K},t) \tag{133}$$

The appropriate boundary condition is $F_s(\underset{\sim}{K},0) = 1$, and Equation 133 yields

$$F_S(\underset{\sim}{K},t) = e^{i\underset{\sim}{K}\cdot\underset{\sim}{v}_d t} e^{-D_T K^2 t} \tag{134}$$

This result is as simple as could be desired since the diffusion term from Equation 103 and the flow term from Equation 125 enter as independent factors.

In place of Equation 125 the first order correlation function for forced diffusion is

$$G^{(1)}(\tau) = \langle I_S \rangle e^{i\underset{\sim}{K}\cdot\underset{\sim}{v}_d \tau} e^{-D_T K^2 \tau} e^{-i\omega_0 \tau} \tag{135}$$

and for the heterodyne experiment Equations 111 and 135 give

$$G^{(2)}_{het}(\tau) = I_L^2 + 2 I_L \langle I_S \rangle \cos(\underset{\sim}{K} \cdot \underset{\sim}{v}_d \tau)\, e^{-D_T K^2 \tau} \tag{136}$$

The second term on the rhs is oscillatory, but decays to zero because of the damping factor $\exp(-D_T K^2 \tau)$. This behavior is illustrated in Figure 14a where $G_{het}(\tau)$ is plotted vs. τ for $\underset{\sim}{K} \cdot \underset{\sim}{v}_d = 100\pi$ and $D_T K^2 = 5\pi$ which correspond to 50 Hz and 2.5 Hz, respectively. By comparison with the previous results for diffusion and flow in heterodyne experiments, we expect the power spectrum associated with Equation 136 to exhibit a line having the frequency shift $\underset{\sim}{K} \cdot \underset{\sim}{v}_d$ and the width (HWHH) $D_T K^2$. The calculation of $I_{het}(\omega)$, which bears this out, proceeds using Equation 92.

$$I^{(2)}_{het}(\omega) = I_L^2\, \delta(\omega) + \frac{2}{\pi} I_L \langle I_S \rangle \operatorname{Re} \int_0^\infty \frac{(e^{i\underset{\sim}{K} \cdot \underset{\sim}{v}_d \tau} + e^{-i\underset{\sim}{K} \cdot \underset{\sim}{v}_d \tau})}{2}\, e^{-D_T K^2 \tau}\, e^{i\omega\tau}\, d\tau \tag{137}$$

The second term on the rhs of Equation 137 is easily evaluated to obtain

$$\frac{I_L \langle I_S \rangle}{\pi} \left[\frac{DK^2}{(DK^2)^2 + (\omega + \underset{\sim}{K} \cdot \underset{\sim}{v}_d)^2} + \frac{DK^2}{(DK^2)^2 + (\omega - \underset{\sim}{K} \cdot \underset{\sim}{v}_d)^2} \right]$$

The function $I^{(2)}_{het}(\omega)$ is illustrated in Figure 14b for the same parameters which were used in part a. Equation 137 is very important since it shows that particles, e.g., molecules or cells, can be distinguished on the basis of their drift velocities, and that very small velocities can be measured.

3. Electrophoretic Light Scattering

The use of quasi-elastic light scattering to study molecules moving in an externally applied DC electric field is perhaps the most significant application of laser velocimetry in biology.[64] The basis of this experiment is that a charged molecule in an electrophoresis cell attains a characteristic drift velocity $\underset{\sim}{v}_d$ which is given by

$$\underset{\sim}{v}_d = u\, \underset{\sim}{E}_{dc} \tag{138}$$

where u is the electrophoretic mobility, characteristic of the molecule in a particular environment, and E_{dc} is the amplitude of the applied field. From simple arguments, similar to those used in Appendix H, the mobility is expected to be proportional to the net charge Ze of the particle and inversely proportional to the friction coefficient f_T. However, the exact dependence of u on these quantities is very complicated for charged macromolecules in electrolyte solutions. In general for spherical particles we can write[65]

$$u = \frac{Ze}{f_T} H(I,T,a) \tag{139}$$

where H is a screening function which depends at least on the ionic strength I, the temperature T, and the particle radius a. An approximate analytic equation is presented below. However, in discussing the experiment it suffices to say that the mobilities for particles of interest, i.e., macro-ions, cells, and viruses, usually lie in the range 1 to 5 μm $s^{-1}V^{-1}$cm. With electric fields in the neighborhood of 10^2 V cm^{-1}, which are easily obtained, the frequency shifts are in the range 10 to 100 Hz for the scattering angle $\theta_s \sim 5°$.

The experimental geometry is shown in Figure 15. The scattering plane is horizontal, i.e., the xy-plane, and the DC electric field is in the +x-direction. According to Equation 137, the frequency shift in radians/s is given by

$$\underset{\sim}{K} \cdot \underset{\sim}{v}_d = 2 k_i v_d \sin(\theta_s/2) \cos\phi = -k_i v_d \sin\theta_s \tag{140}$$

where ϕ is the angle between $\underset{\sim}{K}$ and $\underset{\sim}{v}_d$. Thus, the shift increases as sin θ_s increases, and it might appear that large scattering angles are desirable; however, this is not the case since the linewidth increases as $\sin^2(\theta_s/2)$. The important quantity in this experiment is the resolution which is determined by the ratio of the shift to the linewidth.

$$\left|\frac{\underset{\sim}{K} \cdot \underset{\sim}{v}_d}{D_T K^2}\right| = \frac{k_i u E_{dc} \sin\theta_s}{D_T (2 k_i)^2 [\sin(\theta_s/2)]^2} = \frac{\lambda_0 u E_{dc}}{4\pi n D_T}\left[\frac{\cos(\theta_s/2)}{\sin(\theta_s/2)}\right] \tag{141}$$

For small scattering angles $\cos(\theta_s/2) \sim 1$ and $\sin(\theta_s/2) \sim \theta_s/2$. The resolution is then given by

$$\text{resolution} = \frac{\lambda_0 u E_{dc}}{2\pi n D_T \theta_s} \tag{142}$$

It turns out that small scattering angles are desirable from the standpoint of resolution. Unfortunately, because of problems with stray light and the necessity of using finite collection apertures, about 1° is the practical lower limit on θ_s. An electrophoretic light scattering (ELS) spectrum taken from the work of Hass and Ware on carboxyhemoglobin is shown in Figure 16 to illustrate what can be done.[66] The scattering angle was

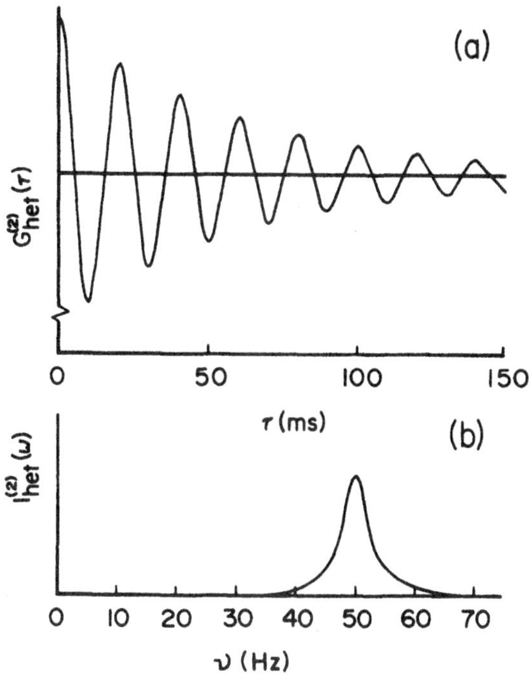

FIGURE 14. (a) Correlation function $G_{het}^{(2)}(\tau)$ calculated using Equation 136, (b) Power spectrum $I_{het}^{(2)}(\omega)$ calculated using Equation 137. The parameters used were $\underset{\sim}{K}\cdot\underset{\sim}{v}_d = 100\pi$ and $D_T K^2 = 5\pi$.

$$\underset{\sim}{k}_i = k_i \hat{j} \quad ; \quad \underset{\sim}{k}_s = k_i(\sin\Theta_s \hat{i} + \cos\Theta_s \hat{j})$$
$$\underset{\sim}{v}_d = v_d \hat{i} \quad ; \quad \underset{\sim}{K} = \underset{\sim}{k}_i - \underset{\sim}{k}_s$$

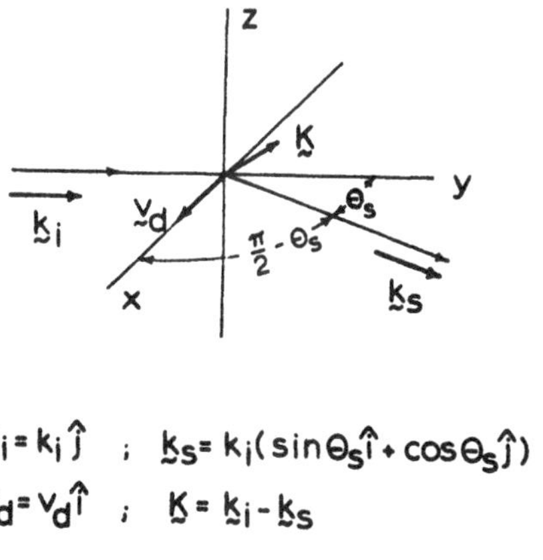

FIGURE 15. The geometry of an experiment in laser velocimetry with $\underset{\sim}{v}_d$ in the scattering plane.

4.18°, the electric field 88.8 V cm^{-1}, the ionic strength of 0.01 M, and the pH 9.5. From the peak position (shift) the electrophoretic mobility was calculated to be u_w^{20} = 2.74 μm s^{-1} V^{-1} cm. The experimental halfwidth (HWHH) was 11.5 ± 0.4 Hz, which is about 1 Hz larger than expected on the basis of measurements of D_T. The discrepancy was attributed to some experimental broadening effect.

ELS is an exciting development which is certain to be increasingly important in biological applications. The number of applications has thus far been surprisingly small. This probably results from technical problems which make the experiment far from routine with present techniques. For example, electrical currents passing through the solution lead to joule heating effects and to concentration polarization. Both of these effects can be reduced by turning the DC electric field on for only a fraction of the time and by alternating the polarity. With cell designs that place charged walls in contact with the scattering sample, electro-osmosis can also be a problem since it causes the solvent to flow. Special coatings on walls and windows can minimize this effect. These and other problems have motivated ingenious cell designs.[66,67] Even when the major problems have been solved there is still a limitation on the ionic strength which can be used, the typical value used being about 10^{-2} *M*.

The heterodyne experiment, which is normally used, also requires careful adjustment to get good mixing efficiency of the scattered and reference beams on the PMT. It should be mentioned that ELS can be done with a homodyne experiment if the laser cross-beam arrangement is used.[68] In this experiment the laser beam is split and then the two resulting beams are crossed in the sample with an intersection angle of about 5°. One interpretation of this experiment is that the crossed beams create a fringe intensity pattern and the scattering particles move through this pattern. Fluctuations in the scattered intensity are thus related to the separation of the fringes and the velocity of the particles. For an analysis of this experiment the reader is referred to the literature on laser anemometry.[69]

The ELS experiment is important because it permits several species in a sample to be resolved and studied simultaneously, and because it permits the electric mobilities to be determined. The mobilities are of interest since they are related to the net charges of the scattering particles. As previously mentioned, the relationship between the mobility and the charge is not simple. An approximate equation derived by Henry is often used for solutions having low ionic strengths. According to Henry's equation

$$u = \frac{Ze}{6\pi\eta a}\left[\frac{X_1(\kappa a)}{1 + \kappa a}\right] \tag{143}$$

where $X_1(\kappa a)$ is a complicated function which has the value 1.5 for $\kappa a >> 1$, monotonically decreases as κa decreases, and approaches 1.0 for $\kappa a << 1$. The parameter κ is the well-known Debye-Hückel constant

$$\kappa = \left[\frac{2N_A e^2 I}{10^3 \epsilon_0 k_e k_B T}\right]^{1/2} \tag{144}$$

The limitations of Henry's equation are discussed by Tanford and in specialized texts.[70]

Rapid *in situ* determination of electrophoretic mobility in a range of ionic strengths is of particular importance in the study of surface charges of cells. Table 4 gives a survey of the types of studies undertaken thus far. An example of the resolution of cell populations by means of ELS is shown in Figure 17 for lymphocytes and erythrocytes.[71] In this work an isoosmotic sucrose buffer solution of pH 7.3 and ionic strength

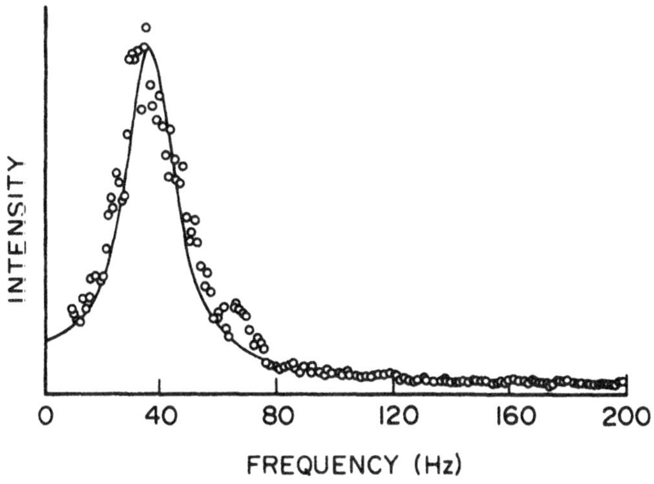

FIGURE 16. Electrophoretic light scattering spectrum of carboxyhemoglobin, 100 μM in heme. (From Haas, D. D. and Ware, B. R., *Anal. Biochem.*, 74, 175, 1976. With permission.)

I = 0.005 was used. The electrodes were platinum sheets, which had been platinized by electrodisposition and coated with bovine serium albumin to prevent reaction with the medium. The mobilities, calculated from the frequency shifts by means of Equation 140, had estimated accuracies of better than 5%, and were reproducible to better than 2% for different donors. The low ionic strength was selected in order to reduce heating and polarization effects, but it had the added advantage of giving larger cell mobilities and better resolution than that found at higher ionic strengths.

In another interesting study, mobilities for virus particles were used by Rimai et al. to calculate surface charge densities σ by means of the identity $Ze = \sigma(4\pi a^2)$.[76] For example with murine leukemia virus (MuLV), where the radius a = 72 nm was determined from diffusion measurements, Equation 143 was used to calculate a surface charge of 7.20×10^{-3} C.m^{-2} at pH 9 and 4.14×10^{-3} C.m^{-2} at pH 5. The value of $X_1(\kappa a)$ required in this calculation was obtained from Reference 1. The mobilities of all of the viruses studied were found to be quite similar in the high pH region.

Finally, we call attention to the study by Smith et al. of the electrophoretic mobilities of human peripheral blood lymphoblasts from normal donors and from patients with acute lymphocytic leukemia.[74] This study was motivated by the idea that surface charge density can be used to distinguish functionally different cells. An ELS cell similar to that described in Reference 66 was used in this work, and the individual averaged spectra required 1 to 5 min to collect. The major conclusion was that the mobilities corresponding to intensity maxima in ELS spectra were 7 to 28% lower for cells from nine leukemic patients than for normal cells. The standard deviation in the mobilities of the normal cells was only 2%. It was also confirmed that cryopreservation had no significant effect on the electrophoretic distribution. The potential clinical applications of this work are obvious.

D. Rotational Motion

1. Anisotropic Molecules

For molecules which are optically anisotropic it is possible to measure the rate of rotational motion by means of light scattering. The rotational diffusion coefficient D_R

Table 4
ELECTROPHORETIC MOBILITIES MEASURED BY MEANS OF ELECTROPHORETIC LIGHT SCATTERING

Macromolecule	Conditions	μ(μm s^{-1} V^{-1} cm)	Ref.
Bovine serum albumin (BSA)	pH 9.55; I = 0.004	2.5	72
BSA dimers		1.7	72
Erythrocytes	0.29 *M* sucrose; 23.5°C	2.84	71
	0.145 *M* sucrose	3.40	71
	pH 7.2, 0.015 *M* NaCl; 4% sorbitol	1.10 ± 0.06	73
Lymphocytes	0.29 *M* sucrose, 23.5°C	2.4	71
	0.145 *M* sucrose	2.75	71
Normal lymphocytes	pH 7.1; 0.0145 *M* NaCl; 4% sorbitol	2.53 ± 0.05	74
Leukemic samples		2.13 ± 0.18	74
Rat thymus lymphocytes (RTL)		3.1	75
RTL stimulated with pokeweed mitogen		1.4	75
Calf thymus DNA	I = 0.004, 20°C	5.9	76
	0.01	5.0	
	0.02	4.4	
	0.05	3.3	
	0.1	2.8	
Tobacco mosaic virus (TMV)	I = 0.001, 20°C	5.2	76
	0.01	3.5	
	0.02	2.7	
	0.05	1.9	
	0.1	1.5	
Murine leukemia virus (MuLV)	pH 9, I = 0.005	2.8 ± 0.1	76
	7	2.7 ± 0.1	
	5	1.65 ± 0.1	
	3	1.02 ± 0.1	
Avian myeloblastosis virus (AMV)	pH 9, I = 0.005	2.64 ± 0.1	76
	7	2.78 ± 0.2	
	5	1.84 ± 0.05	
	3	1.63 ± 0.05	
Feline leukemia virus (FeLV)	pH 9, I = 0.005	3.19 ± 0.2	76
	7	2.71 ± 0.1	
	5	2.20 ± 0.1	
	3	2.60 ± 0.2	
Murine mammary tumor virus (MuMTV)	pH 9; I = 0.005	3.01 ± 0.2	76
	7	2.68 ± 0.2	
	5	2.36 ± 0.2	
Hamster fibroblasts	I = 0.15	1.0	76

often provides an adequate description of this motion for macromolecules. One reason for interest in D_R is that it is a sensitive function of the molecular radius. In 1928 Debye showed that[77]

$$D_R = \frac{k_B T}{f_R} \tag{145}$$

where f_R is the rotational friction coefficient. Stokes had previously shown that for a

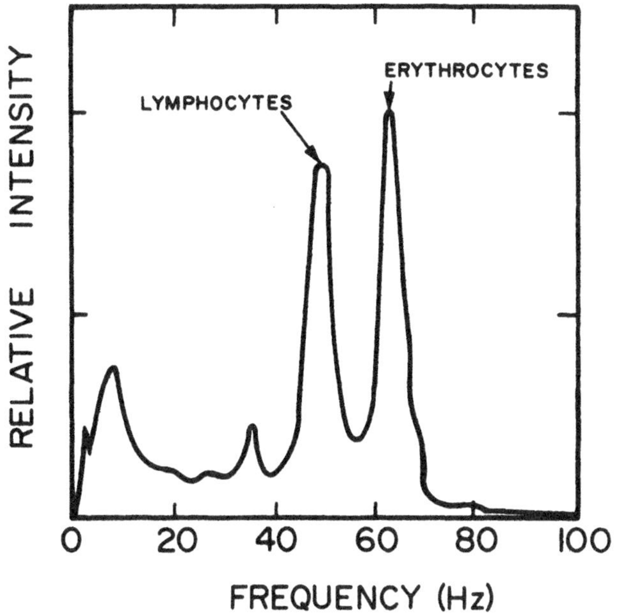

FIGURE 17. Electrophoretic light scattering spectrum of a mixture of erythrocytes and lymphocytes in 0.29 *M* sucrose buffer of ionic strength 0.005. The scattering angle was $\theta_s = 11.5°$. (From Uzgiris, E. E. and Kaplan, J. H., *Anal. Biochem.*, 60, 455, 1974. With permission.)

spherical particle of radius a in a solvent having the viscosity η, $f_R = 8\pi\eta a^3$. Thus D_R is seen to depend on the inverse cube of the radius.

A quantitative description of the scattering experiment requires that we specify the polarizability tensor for the scattering particle (see Appendix A). For simplicity we assume axial symmetry so that in the principal axes system the polarizability tensor has the form

$$\underset{\approx}{\alpha} = \begin{pmatrix} \alpha_\perp & 0 & 0 \\ 0 & \alpha_\perp & 0 \\ 0 & 0 & \alpha_\parallel \end{pmatrix} \tag{146}$$

The molecular orientation is shown in Figure 18 where x, y, and z specify the laboratory fixed coordinate system, equivalent to that in Figure 1, and X, Y, and Z are the principal axes for the polarizability tensor. We take the unique axis to be Z, the orientation of which is described by the polar angles θ and ϕ with respect to the laboratory frame. The orientation of X is arbitrary, because of the axial symmetry, however, we choose to place X in the plane which contains both the z and Z axes.

We assume that the incident light propagates in the + y-direction and that it is polarized in the z-direction, i.e., the incident light is vertically polarized. The scattered light is collected at the scattering angle θ_s in the xy plane; and we are concerned with

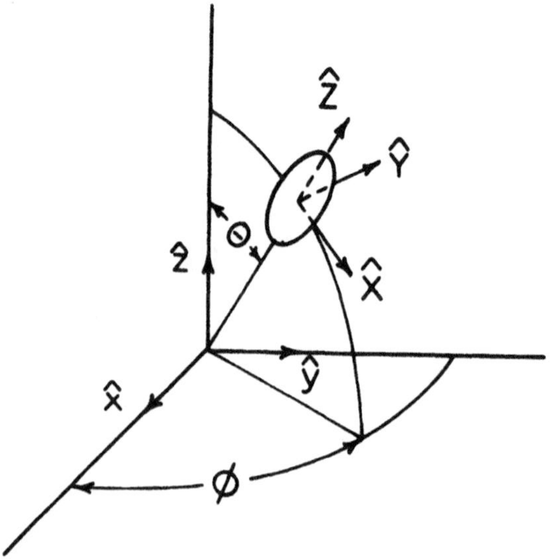

FIGURE 18. The laboratory coordinate system (xyz) and the principal axes system (XYZ) for a molecule having axial symmetry.

both the vertically and horizontally polarized components, the intensities of which are denoted by I_{VV} and I_{VH}, respectively. The induced dipole moments p_z and p_y are responsible for the scattered intensities, and according to Equation 210 these moments depend on the components α_{zz} and α_{yz}, respectively, of the polarizability tensor. For example $p_y = \alpha_{yz} E_z$. Our task then is to express α_{zz} and α_{yz} in terms of $\alpha_\perp$, $\alpha_\parallel$, and the molecular orientation. The transformation of $\underset{\approx}{\alpha}$ from the principal axes system to the laboratory frame is easily accomplished if the unit vectors $\hat{x}$, $\hat{y}$, and $\hat{z}$ are expressed in terms of the unit vectors, $\hat{X}$, $\hat{Y}$, and $\hat{Z}$.

By inspection of Figure 18 we see that

$$\hat{X} = \cos\theta\cos\phi\,\hat{x} + \cos\theta\sin\phi\,\hat{y} - \sin\theta\,\hat{z} \tag{147a}$$

$$\hat{Y} = -\sin\phi\,\hat{x} + \cos\phi\,\hat{y} \tag{147b}$$

$$\hat{Z} = \sin\theta\cos\phi\,\hat{x} + \sin\theta\sin\phi\,\hat{y} + \cos\theta\,\hat{z} \tag{147c}$$

The inverse of Equation 147, while not as easy to visualize, can be obtained by transposing the coefficients. Thus

$$\hat{x} = \cos\theta\cos\phi\,\hat{X} - \sin\phi\,\hat{Y} + \sin\theta\cos\phi\,\hat{Z} \tag{148a}$$

$$\hat{y} = \cos\theta\sin\phi\,\hat{X} + \cos\phi\,\hat{Y} + \sin\theta\sin\phi\,\hat{Z} \tag{148b}$$

$$\hat{z} = -\sin\theta\,\hat{X} + \cos\theta\,\hat{Z} \tag{148c}$$

Now armed with both $\underset{\approx}{\alpha}$ and the unit vectors, $\hat{x}$, $\hat{y}$, and $\hat{z}$ referred to the principal axes system, we can project out the desired components.

$$\alpha_{zz} = \hat{z}\cdot\underset{\approx}{\alpha}\cdot\hat{z}$$

$$\alpha_{zz} = (-\sin\theta, 0, \cos\theta)\begin{pmatrix}\alpha_\perp & 0 & 0\\ 0 & \alpha_\perp & 0\\ 0 & 0 & \alpha_\parallel\end{pmatrix}\begin{pmatrix}-\sin\theta\\ 0\\ \cos\theta\end{pmatrix} \tag{149}$$

$$\alpha_{zz} = \alpha_\perp \sin^2\theta + \alpha_\parallel \cos^2\theta$$

Similarly

$$\alpha_{yz} = \hat{y}\cdot\underset{\approx}{\alpha}\cdot\hat{z}$$
$$= (\alpha_\parallel - \alpha_\perp)\sin\theta\cos\theta\sin\phi \tag{150}$$

These equations are special cases of the application of Equation 212.

To see how α_{zz} and α_{yz} fit into the scattering theory, we return to Equation 83 where the scattering amplitude for the ℓth particle is specified by A_ℓ. In the VV experiment A_ℓ is proportional to α_{zz} and Equation 83 becomes

$$G^{(1)}_{VV}(\tau) = AN <\alpha^*_{zz}(0)\ \alpha_{zz}(\tau)> F_s(\underset{\sim}{K},\tau)\, e^{-i\omega_0\tau} \tag{151}$$

while in the VH experiment A_ℓ is proportional to α_{yz} and

$$G^{(1)}_{VH}(\tau) = AN<\alpha^*_{yz}(0)\ \alpha_{yz}(\tau)> F_s(\underset{\sim}{K},\tau)\, e^{-i\omega_0\tau} \tag{152}$$

Here A is a constant which includes E_o^2 and all of the other quantities from Equation 8 which determine the intensity. The evaluation of the correlation functions of elements of the polarizability tensor must be based on a particular model of the motion. In dealing with rotational diffusion it turns out to be much more convenient to express the angular dependences of α_{zz} and α_{yz} in terms of the spherical harmonics $Y_{\ell m}(\theta,\phi)$ rather than the trigonometric functions of Equations 149 and 150. The functions which appear in transformations of the polarizability tensor, i.e., spherical harmonics having $\ell = 2$, are listed in Table 5. By solving for $\cos^2\theta$ and $\sin^2\theta$ in terms of $Y_{2,0}(\theta,\phi)$ it is easy to show that

$$\alpha_{zz} = \alpha + \beta\sqrt{\frac{16\pi}{45}}\ Y_{2,0}(\theta,\phi) \tag{153}$$

Table 5

SPHERICAL HARMONICS FOR $\ell = 2$

$$m = 2 \quad Y_{2,2}(\theta,\phi) = +\sqrt{\frac{15}{32\pi}} \sin^2\theta\, e^{2i\phi}$$

$$m = 1 \quad Y_{2,1}(\theta,\phi) = -\sqrt{\frac{15}{8\pi}} \sin\theta\cos\theta\, e^{i\phi}$$

$$m = 0 \quad Y_{2,0}(\theta,\phi) = +\sqrt{\frac{5}{16\pi}} (3\cos^2\theta - 1)$$

$$m = -1 \quad Y_{2,-1}(\theta,\phi) = +\sqrt{\frac{15}{8\pi}} \sin\theta\cos\theta\, e^{-i\phi}$$

$$m = -2 \quad Y_{2,-2}(\theta,\phi) = +\sqrt{\frac{15}{32\pi}} \sin^2\theta\, e^{-2i\phi}$$

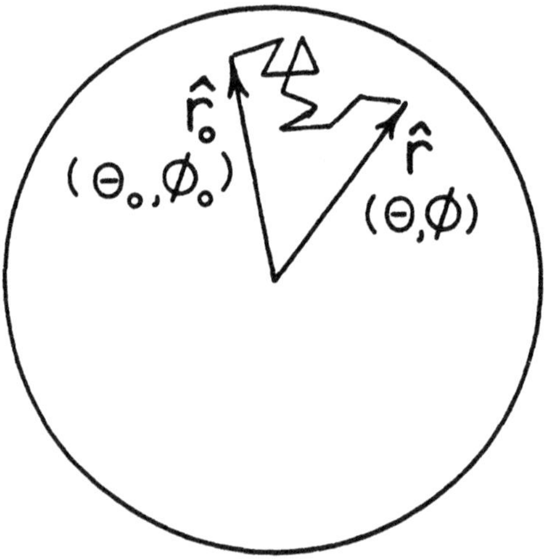

FIGURE 19. Rotational diffusion of the unit vector $\hat{r}$. The t = O orientation is specified by $\hat{r}_o$.

Also, the combination $Y_{2,-1}(\theta,\phi) + Y_{2,1}(\theta,\phi)$ when used with Euler's theorem immediately gives

$$\alpha_{yz} = i\beta\sqrt{\frac{2\pi}{15}}\,[Y_{2,1}(\theta,\phi) + Y_{2,-1}(\theta,\phi)] \tag{154}$$

Here $\alpha = (\alpha_{\parallel} + 2\alpha_{\perp})/3$ and $\beta = (\alpha_{\parallel} - \alpha_{\perp})$, which are consistent with Equations 213 and 214. When Equations 153 and 154 are combined with Equations 151 and 152, it is clear that we must evaluate correlation functions of the form $\langle Y^*_{\ell m}(\theta_o,\phi_o)$

$Y_{\ell'm'}(\theta,\phi)\rangle$ where (θ_o,ϕ_o) and (θ,ϕ) specify the molecular orientations at $t = 0$ and $t = \tau$, respectively.

2. Rotational Diffusion

In rotational diffusion as described by Debye, a particle immersed in a viscous fluid is assumed to undergo numerous collisions with the solvent molecules. These collisions cause the particle to rotate through a sequence of small angular steps in random directions. If $\hat{r}$ is a unit vector fixed in the particle, we imagine that the tip of $\hat{r}$ undergoes a random walk on the surface of the unit sphere as shown in Figure 19.

The following probability functions are useful in describing the time evolution of $\hat{r}$ and functions of $\hat{r}$.

1. $P(\hat{r}_o) = P(\theta_o,\phi_o)$ = probability that the orientation is (θ_o,ϕ_o) at $t = 0$.
2. $P(\hat{r}_o|\hat{r},\tau) = P(\theta_o,\phi_o|\theta,\phi,\tau)$ = conditional probability that the orientation is (θ,ϕ) at $t = \tau$ given that it was (θ_o,ϕ_o) at $t = 0$.
3. $P(\hat{r}_o)\ P(\hat{r}_o|\hat{r},\tau)$ = probability that orientation is (θ_o,ϕ_o) at $t = 0$ and (θ,ϕ) at $t = \tau$.

In calculating the correlation functions for the spherical harmonics $Y_{\ell m}(\theta,\phi)$ the function $P(\hat{r}_o)\ P(\hat{r}_o|\hat{r},\tau)$ is required. The first factor, $P(\hat{r}_o)$, is a constant since all initial orientations are equally likely. The probability that there is some initial orientation is, of course, unity. Thus

$$\int P(\hat{r}_o)\,d\Omega_o = P(\theta_o,\phi_o)\int_{\phi_o=O}^{2\pi}\int_{\theta_o=O}^{\pi} \sin\theta_o\,d\theta_o\,d\phi_o = 1 \tag{155}$$

and $P(\theta_o,\phi_o) = 1/4\pi$.

The second factor, $P(\hat{r}_o|\hat{r},\tau)$, obeys the rotational diffusion equation

$$\frac{\partial}{\partial\tau}P(\hat{r}_o|\hat{r},\tau) = D_R\,\nabla^2 P(\hat{r}_o|\hat{r},\tau) \tag{156}$$

This equation and its solutions are discussed in Appendix I where it is shown that

$$P(\hat{r}_o|\hat{r},\tau) = \sum_{\ell=O}^{\infty}\sum_{m=-\ell}^{+\ell} Y_{\ell m}(\theta_o,\phi_o)\,Y^*_{\ell m}(\theta,\phi)\,e^{-\ell(\ell+1)D_R\tau} \tag{157}$$

If $\hat{r}$ is identified with $\hat{Z}$ for an axially symmetric macromolecule, the correlation functions required in Equations 151 and 152 contain terms of the form

$$\langle Y^*_{\ell m}(\theta_o,\phi_o)\,Y_{\ell'm'}(\theta,\phi)\rangle = \iint Y^*_{\ell m}(\hat{r}_o)\,Y_{\ell'm'}(\hat{r})\,P(\hat{r}_o)\,P(\hat{r}_o|\underset{\sim}{\hat{r}},\tau)\,d\Omega_o\,d\Omega \tag{158}$$

Therefore, using Equations 151 and 157 the following integral must be evaluated

$$\iint Y^*_{\ell m}(\hat{r}_o)\, Y_{\ell' m'}(\hat{r}) \frac{1}{4\pi} \sum_{\ell'', m''} Y^*_{\ell'' m''}(\hat{r}_o)\, Y_{\ell'' m''}(\hat{r})\, e^{-\ell''(\ell''+1) D_R \tau}\, d\Omega_o d\Omega$$

where for simplicity we have used $\hat{r}_o$ and $\hat{r}$ to specify the orientations (θ_o, ϕ_o) and (θ, ϕ), respectively. This integral is easily evaluated using the special property of spherical harmonics, namely

$$\int_{\phi=0}^{2\pi} \int_{\theta=0}^{\pi} Y^*_{\ell m}(\theta,\phi)\, Y_{\ell' m'}(\theta,\phi) \sin\theta \, d\theta \, d\phi = \delta_{\ell\ell'}\, \delta_{mm'} \tag{159}$$

The Kronecker delta δ_{ij} is defined so that $\delta_{ij} = 1$ for $i = j$ and $\delta_{ij} = 0$ for $i \neq j$. It is this property of spherical harmonics which justifies the introduction of the $Y_{\ell m}(\theta,\phi)$ in Equations 153 and 154. Equation 159 is used in both the integrations over (θ_o, ϕ_o) and (θ, ϕ) in 158 to obtain

$$\langle Y^*_{\ell m}(\theta_o, \phi_o)\, Y_{\ell' m'}(\theta,\phi) \rangle = \frac{1}{4\pi}\, \delta_{\ell\ell'}\, \delta_{mm'}\, e^{-\ell(\ell+1) D_R \tau} \tag{160}$$

We are now in a position to express $G^{(1)}_{VV}(\tau)$ and $G^{(1)}_{VH}(\tau)$ in terms of the diffusion coefficients D_R and D_T. First Equations 153, 154, and 160 are combined to obtain

$$\begin{aligned} \langle \alpha^*_{zz}(0)\, \alpha_{zz}(\tau) \rangle &= \left\langle \left[\alpha + \beta \sqrt{\frac{16\pi}{45}}\, Y^*_{2,0}(\theta_o,\phi_o)\right]\left[\alpha + \beta \sqrt{\frac{16\pi}{45}}\, Y_{2,0}(\theta,\phi)\right] \right\rangle \\ &= \alpha^2 + \frac{4}{45} \beta^2 e^{-6 D_R \tau} \end{aligned} \tag{161}$$

and

$$\begin{aligned} \langle \alpha^*_{yz}(0)\, \alpha_{yz}(\tau) \rangle &= \frac{2\pi}{15} \beta^2 \left\langle \left[Y^*_{2,1}(\theta_o,\phi_o) + Y^*_{2,-1}(\theta_o,\phi_o)\right]\left[Y_{2,1}(\theta,\phi) + Y_{2,-1}(\theta,\phi)\right] \right\rangle \\ &= \frac{1}{15} \beta^2 e^{-6 D_R \tau} \end{aligned} \tag{162}$$

The first-order correlation functions can then be written as

$$G_{VV}^{(1)}(\tau) = AN\left[\alpha^2 + \frac{4}{45}\beta^2 e^{-6D_R\tau}\right] e^{-D_TK^2\tau} e^{-i\omega_0\tau} \tag{163}$$

$$G_{VH}^{(1)}(\tau) = AN\frac{\beta^2}{15} e^{-6D_R\tau} e^{-D_TK^2\tau} e^{-i\omega_0\tau} \tag{164}$$

In the VV experiment the correlation function contains an "isotropic" term identical to that found for spherical particles, but in addition contains a term, the amplitude of which depends on the anisotropy of the polarizability. The latter term decays with the time constant $\tau_c = (6D_R + D_TK^2)^{-1}$. In contrast to this, $G_{VH}^{(1)}(\tau)$ only contains an "anisotropic" term. It should be recalled that K is proportional to $\sin(\theta_s/2)$ and that the dependence on D_T can in principle be removed by extrapolating the decay rate to $\theta_s = O$. In a homodyne experiment of the VV type the relevant equation, obtained through use of Equation 163 and the Siegert relation, is

$$g_{VV}^{(2)}(\tau) = 1 + \frac{\left[B_0^2 e^{-2D_TK^2\tau} + 2B_0B_2 e^{-(2D_TK^2 + 6D_R)\tau} + B_2^2 e^{-2(D_TK^2 + 6D_R)\tau}\right]}{(B_0 + B_2)^2} \tag{165}$$

where $B_0 = \alpha^2$ and $B_2 = 4\beta^2/45$.

The optical spectra associated with $G_{VV}^{(1)}(\tau)$ and $G_{VH}^{(1)}(\tau)$ are also of interest. The application of Equation 93 gives

$$\begin{aligned} I_{VV}^{(1)}(\omega) &= \frac{<I^{ISO}>}{\pi}\left[\frac{D_TK^2}{(D_TK^2)^2 + (\omega_0 - \omega)^2}\right] \\ &+ \frac{4}{3}\frac{<I^{ANISO}>}{\pi}\left[\frac{6D_R + D_TK^2}{(6D_R + D_TK^2)^2 + (\omega_0 - \omega)^2}\right] \end{aligned} \tag{166}$$

$$I_{VH}^{(1)}(\omega) = \frac{<I^{ANISO}>}{\pi}\left[\frac{6D_R + D_TK^2}{(6D_R + D_TK^2)^2 + (\omega_0 - \omega)^2}\right] \tag{167}$$

where $<I_s^{ANISO}> = <I_{VH}>$ and $<I_s^{ISO}>$ is the scattering intensity resulting from α^2 alone. Typically $6D_R >> D_TK^2$, and for small globular proteins D_R is so large that the optical linewidths can be measured directly without the use of light-beating spectroscopy. Lysozyme provides a good example. For this molecule $6D_R(20°,w) \simeq 10^8\ s^{-1}$, and for the scattering angle $\theta_s = 90°$ the quantity $D_T(20°,w)\ K^2$ is only about $4 \times 10^4\ s^{-1}$. The HWHH for the depolarized spectrum is between 10 and 20 MHz, which can be measured using a Fabry-Perot interferometer. Dubin et al. used this method to obtain $D_R(20°,w) = 16.7 \times 10^6\ s^{-1}$.[51] By assuming that lysozyme can be approximated as a prolate ellipsoid of revolution, they were also able to use Perrin's expressions for f_T and f_R in terms of the ratio of the major to minor axes to show that the major and minor axes are (55 ± 1)Å and (33 ± 1)Å, respectively, for the hydrated molecule. For an ellipsoid having the semiaxes a,b, and b, the required expression for f_T is given in Appendix H. Perrin's equation for f_R for rotation of the symmetry axis is[78,79]

$$f_R = \frac{16\pi\eta a^3}{3}\left\{\frac{1-(b/a)^2}{[2-(b/a)^2]\,G(b/a)-1}\right\} \tag{168}$$

where G(b/a) was defined in connection with f_T in Appendix H.

Another simple case, which is important, arises for molecules which can be approximated as long thin rods of radius b and length L. If $20L \geqslant \lambda$, the structure factor $S(\underset{\sim}{K})$ introduces an intensity fluctuation even if the molecule is composed of isotropic segments. The form of Equation 165 is approximately correct for KL<8, but the magnitudes of B_o and B_2 must be looked up in tables.[80] For larger values of KL additional terms must be included. A well-studied example of a rod-shaped molecule is tobacco mosaic virus (TMV) for which $D_R \simeq 320\ s^{-1}$.[81] The linewidth corresponding to this value of the rotational diffusion coefficient is, of course, so small that light beating spectroscopy must be used.

E. Motility

Microorganisms often persist in their translational motion for distances which are much greater than $|\underset{\sim}{K}|^{-1}$. Except for the distribution of velocities, the situation is similar to that in a low pressure gas. The velocity distribution function $P(\underset{\sim}{v})$ is, in fact, the quantity of interest. The determination of $P(\underset{\sim}{v})$ by means of quasi-elastic light scattering was demonstrated by Nossal in 1971.[82] To show how this is done, we write Equation 82 as follows:

$$G^{(1)}(\tau) = \sum_{\ell=1}^{N} \langle A_\ell^*(O)\, A_\ell(\tau)\rangle \langle e^{i\underset{\sim}{K}\cdot[\underset{\sim}{r}_\ell(\tau) - \underset{\sim}{r}_\ell(O)]}\rangle E_o^2 e^{-i\omega_o\tau} \tag{169}$$

Thus far we have assumed that rotational and translational motions are independent and that the motions of different scattering particles are not correlated. If in addition we assume that either the particles are small and isotropic or that for large particles the scattering angle is small, then $\langle A^*_\ell(O)\, A_\ell(\tau)\rangle = \langle|A_\ell(O)|^2\rangle$ and Equation 169 can be written as

$$G^{(1)}(\tau) = \frac{\langle I_S\rangle}{N} \sum_{\ell=1}^{N} \langle e^{i\underset{\sim}{K}\cdot[\underset{\sim}{r}_\ell(\tau) - \underset{\sim}{r}_\ell(O)]}\rangle\, e^{-i\omega_o\tau} \tag{170}$$

The assumption of linear trajectories permits $r_\ell(\tau) - r_\ell(O)$ to be replaced with $v_\ell\tau$ where v_ℓ is the velocity for the ℓth particle or microorganism. Since the velocity distribution is isotropic, we can write

$$F_S(\underset{\sim}{K},\tau) = N^{-1} \sum_{\ell=1}^{N} \langle e^{i\underset{\sim}{K}\cdot\underset{\sim}{v}_\ell\tau}\rangle = \int P(v)\, e^{i\underset{\sim}{K}\cdot\underset{\sim}{v}\tau}\, d^3v \tag{171}$$

where P(v) depends only on the magnitude of $\underset{\sim}{v}$. For dilute gases $P(\underset{\sim}{v})$ is the Maxwell-Boltzmann distribution and Equation 171 is easily evaluated.[83]

In general the form of P(v) is unknown and must be obtained from the experimentally determined function $F_s(\underset{\sim}{K},t)$. For the purpose of evaluating the integral in Equation 171, we take the direction of $\underset{\sim}{K}$ to define the polar axis of a spherical polar coordinate system so that $\underset{\sim}{K}\cdot\underset{\sim}{v}\tau = \underset{\sim}{K}v(\cos\alpha)\tau$ where α is the angle between $\underset{\sim}{K}$ and $\underset{\sim}{v}$. Thus

$$
\begin{aligned}
F_S(K,\tau) &= \int_{\phi=0}^{2\pi}\int_{\alpha=0}^{\pi}\int_{v=0}^{\infty} P(v)\, e^{iKv(\cos\alpha)\tau}\, d\phi \sin\alpha d\alpha\, v^2 dv \\
&= 2\pi\int_{v=0}^{\infty} P(v)\left[\int_{\rho=-1}^{1} e^{iKv\rho\tau}\, d\rho\right] v^2 dv\ ;\ \rho = \cos\alpha \\
&= 4\pi\int_{v=0}^{\infty} P(v)\,\frac{\sin(Kv\tau)}{(kv\tau)}\, v^2 dv
\end{aligned}
\tag{172}
$$

If we define the swimming speed distribution $P_s(v) = 4\pi v^2 P(v)$, then Equation 172 shows that $(K\tau)\,F_s(\underset{\sim}{K},\tau)$ is the Fourier sine transform of $P_s(v)/v$, i.e.,

$$
(K\tau)\, F_S(\underset{\sim}{K},\tau) = \int_0^{\infty} \frac{P_S(v)}{v}\, \sin(Kv\tau)\, dv
$$

The inverse transform immediately gives

$$
P_S(v) = \frac{2v}{\pi}\int_0^{\infty} [(K\tau)\, F_S(\underset{\sim}{K},\tau)]\, \sin(Kv\tau)\, d(K\tau) \tag{173}
$$

which is the desired result. The result of applying this equation to an experimentally determined correlation function is shown in Figure 20. For details the reader is referred to the paper by Nossal et al.[84]

The simple treatment described here has been fairly successful for samples containing swimming bacteria. In certain cases corrections are required to take into account (1) distributions of particle sizes, (2) the presence of both motile and nonmotile bacteria, and (3) contributions from rotational motions.[85] Another quite interesting and useful effect is evident in some studies of bacteria. Namely, the number of bacteria (particles) in the scattering volume is so small that number fluctuations become important. The quantity $\langle\delta N(0)\delta N(\tau)\rangle$, which is related to the mean square velocity and the mean free path, can be determined in these cases.[86] The general problem of number fluctuations is discussed in Sections III.F.

F. Number Fluctuations

In deriving an expression for $G^{(2)}(\tau)$, in Section III.A we assumed that the number

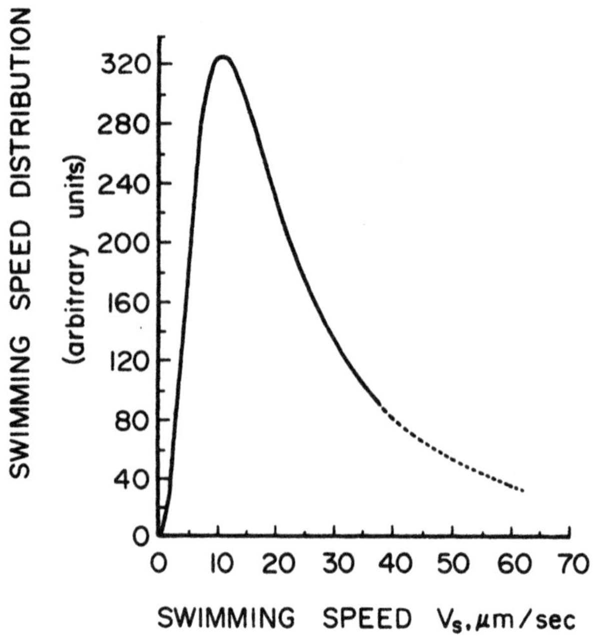

FIGURE 20. The swimming speed distribution $P_s(v)$ for motile bacteria (*Escherichia coli* K_{12}) at T = 25°C calculated from experimental data using Equation 173. (From Nossal, R., Chen, S.-H., and Lai, C.-C., *Opt. Commun.*, 4, 35, 1971. With permission.)

of particles in the scattering volume was $N = \langle N \rangle$. This is usually a very good approximation because N is normally so large that fractional fluctuations are negligible, and when N is small the intensity of the scattered light is very low. However, the situation changes when the scattering particles are very large. A small number of such particles may be in the scattering volume; and, since the scattered intensity is roughly proportional to the square of the mass, the scattered intensity may still be significant. The number of particles in the scattering volume at time t can be written as $N(t) = \langle N \rangle + \delta N(t)$ where $\langle N \rangle$ is the time average value of N(t). In this section we are concerned with the "number fluctuations" δN and their effects in quasi-elastic light scattering (QLS).[87]

Two features of number fluctuations should be noted at the beginning. First, the mean square fluctuations are usually very small. In fact, for an ideal solution $\langle (\delta N)^2 \rangle / \langle N \rangle^2 = \langle (\delta C)^2 \rangle / \langle C \rangle^2 = \langle N \rangle^{-1}$ and the fractional fluctuation in the scattered intensity is expected to roughly equal $\langle N \rangle^{-1/2}$. If $\langle N \rangle = 10^4$, the fluctuation will only be about 1%. The second feature is that the time scale of the fluctuations $\delta N(t)$ is usually very long compared to fluctuations in the phase factors which appear in $G^{(1)}(\tau)$. A fluctuation in N(t) occurs when a particle enters or leaves the scattering volume, and the characteristic time τ_N for the decay of the function $\langle \delta N(O) \delta N(\tau) \rangle$ depends on the dimensions of the scattering volume. For example, in diffusion the mean-square displacement which occurs in the time τ is given by $\langle x^2 \rangle = 2D_T\tau$. Therefore, if the diameter of the scattering volume is of order L, we estimate that $\tau_N \sim L^2/D_T$.

If number fluctuations are taken into account in the expression for E_s, new derivations must be given for $G^{(1)}(\tau)$ and $G^{(2)}(\tau)$. To anticipate the results, we will be able to show that $G^{(1)}(\tau)$ is unaffected by number fluctuations. In contrast to this, $G^{(2)}(\tau)$

is found to depend on $<\delta N(0)\delta N(\tau)>$. The previously stated relationship between $G^{(2)}(\tau)$ and $G^{(1)}(\tau)$ does not hold in this case since the number of particles is too small to permit the Gaussian assumption.

Following Berne and Pecora,[83] we rewrite Equation 77 as

$$E_S(t) = \sum_{j=1} A_j b_j(t) e^{i\underset{\sim}{K}\cdot\underset{\sim}{r}_j} E_0 e^{-i\omega_0 t} \tag{174}$$

where $b_j(t) = 1$ if the jth particle is in the scattering volume at time t and O otherwise. We assume that the scatterers are spherical and that their motions are statistically independent of $b_j(t)$. Equation 82 can now be rewritten as

$$G^{(1)}(\tau) = \sum_{j=1} |A_j|^2 <b_j(O)\, b_j(\tau)> <e^{i\underset{\sim}{K}\cdot[\underset{\sim}{r}_j(\tau) - \underset{\sim}{r}_j(O)]}> E_0^2 e^{-i\omega_0\tau} \tag{175}$$

The crucial step is to realize that $\tau_N >> (D_T K^2)^{-1}$ so that $b_j(\tau)$ can be replaced by $b_j(O)$ without affecting the result, i.e., the factor containing $\underset{\sim}{K}$ decays to zero before $b_j(\tau)$ deviates much from $b_j(O)$ on average. Also, we notice that $b_j^2(O) = b_j(O)$ and that $<\Sigma_{j=1} b_j(O)> = <N(O)>$. The conclusion is that

$$G^{(1)}(\tau) = <N> |A_j|^2 F_S(\underset{\sim}{K},\tau) E_0^2 e^{-i\omega_0\tau} \tag{176}$$

which is equivalent to Equation 94.

Since $G^{(1)}(\tau)$ cannot be derived from $G^{(1)}(\tau)$ by means of the Siegert relation in this case, we must return to the definition of $G^{(2)}(\tau)$. Thus from Equation 87 for the reduced second-order correlation function

$$\begin{aligned} g^{(2)}(\tau) &= <E_S^*(O)\, E_S(O)\, E_S^*(\tau)\, E_S(\tau)> / <I_S>^2 \\ &= \frac{1}{<N>^2} < \sum_{i,j,k,\ell} b_i(O)\, b_j(O)\, b_k(\tau)\, b_\ell(\tau)\, e^{i\underset{\sim}{K}\cdot[\underset{\sim}{r}_j(O) - \underset{\sim}{r}_i(O) + \underset{\sim}{r}_\ell(\tau) - \underset{\sim}{r}_k(\tau)]}> \end{aligned} \tag{177}$$

If any of the terms in this summation have one index which is unique, e.g., i is different from j,k, and ℓ, then the ensemble average of this term vanishes unless $\underset{\sim}{K} = 0$. This occurs because the unique part can be factored out of the term and averaged separately. For example, if j is the unique index in the nth term, then one factor in this term is $<b_j e^{i\underset{\sim}{K}\cdot\underset{\sim}{r}_j}>$. The average here implies an integration over the scattering volume and since

$$v^{-1}\int e^{i\underset{\sim}{K}\cdot\underset{\sim}{r}_j}\, d^3r_j \propto \delta(\underset{\sim}{K})$$

this term only produces forward scattering. Of the terms without unique indexes, there are only two types which can contribute to scattering. First we consider terms having $i=j$ and $k=\ell$ but no other restrictions. From Equation 177 we see that the exponent vanishes and we are left with

$$\frac{1}{<N>^2} \sum_{i,j} <b_i(O)\, b_j(\tau)> \tag{178}$$

where we have used the fact that $b_\ell^2 = b_\ell$. The other type of term which contributes has $i=\ell$ and $j=k$ but $i \neq j$. This gives

$$\frac{1}{<N>^2} \sum_{i \neq j} <b_i(O)\, b_i(\tau)\, b_j(O)\, b_j(\tau)> <e^{i\underset{\sim}{K}\cdot[\underset{\sim}{r}_i(\tau) - \underset{\sim}{r}_i(O)]}> <e^{-i\underset{\sim}{K}\cdot[\underset{\sim}{r}_j(\tau) - \underset{\sim}{r}_j(0)]}> \tag{179}$$

where we have used the fact that in a dilute solution the motion of particle j is uncorrelated with that of particle i. Since the factor containing $r_\ell(\tau) - r_\ell(O)$ decays rapidly to zero, we are justified in replacing $b_i(\tau)$ and $b_j(\tau)$ with $b_i(O)$ and $b_j(O)$, respectively. Finally, using $b_\ell^2 = b_\ell$ this type of term contributes

$$\frac{1}{<N>^2} \sum_{i \neq j} <b_i(O)\, b_j(O)>\, |\, F_S(\underset{\sim}{K},\tau)\, |^2 \tag{180}$$

Consideration of the terms having $i=k$ and $j=\ell$ but $i \neq j$ shows that they contribute only to forward scattering.

The summation which appears in Equation 178 can be written in terms of N as follows:

$$\sum_{i,j} <b_i(O)\, b_j(\tau)> = <\sum_{i=1} b_i(O) \sum_{j=1} b_j(\tau)> = <N(O)\, N(\tau)> \tag{181}$$

and using $N(t) = <N> + \delta N(t)$ this becomes

$$<N(O)N(\tau)> = <N>^2 + <\delta N(0)\, \delta N(\tau)> \tag{182}$$

since $<\delta N(0)> = 0$. The summation in Equation 180 is given by

$$\sum_{i \neq j} <b_i(O) b_j(O)> = <N(N-1)> = <N^2> - <N> \tag{183}$$

The evaluation of the averages requires that the proper distribution function be selected. At low concentrations the probability of finding N particles in the scattering volume is given by the Poisson distribution

$$P(N) = \frac{\langle N\rangle^N e^{-\langle N\rangle}}{N!} \tag{184}$$

Using this function it can be shown that

$$\langle N^2\rangle = \sum_N N^2 P(N) = \langle N\rangle^2 + \langle N\rangle \tag{185}$$

Then combining Equations 181, 183, and 185 we obtain

$$\begin{aligned} g^{(2)}(\tau) &= \frac{1}{\langle N\rangle^2}\left[\langle N\rangle^2 + \langle N\rangle^2 |F_S(\underset{\sim}{K},\tau)|^2 + \langle \delta N(0)\,\delta N(\tau)\rangle\right] \\ &= 1 + |F_S(\underset{\sim}{K},\tau)|^2 + \frac{\langle \delta N(0)\,\delta N(\tau)\rangle}{\langle N\rangle^2} \end{aligned} \tag{186}$$

The first two terms on the rhs of Equation 186 appear in the Siegert relation, which is based on the Gaussian approximation, while the last term is new. At $\tau = 0$, $g^{(2)}(\tau)$ is equal to $2 + \langle N\rangle^{-1}$, then for $\tau > (D_T K^2)^{-1}$ it drops to $1 + \langle N\rangle^{-1}$, and finally for $\tau > \tau_N$ it approaches unity.

Number fluctuations in other types of experiments have been used in recent years to obtain kinetic information from fluctuations about equilibrium. For example fluctuations in the intensity of fluoresence have been analyzed to obtain the rate of binding of the dye ethydium bromide to DNA.[88] Also, fluctuations in conductance have been related to the rates of ionic association reactions.[89] In principal this kind of analysis can be carried out with any type of signal which is proportional to the number of molecules. However, success in a number fluctuation experiment demands a relatively small value of $\langle N\rangle$ and extreme sensitivity in detection. No one has reported measurements of fluctuations in absorption. Of course, these comments refer to fluctuations at thermal equilibrium. Nonequilibrium situations such as turbulance in gases produce much larger fractional fluctuations which can even be detected in Raman spectroscopy.[90]

G. Chemical Reactions

In chemical reactions the products differ from the reactants in their polarizabilities, diffusion coefficients, electrophoretic mobilities, and perhaps other properties. It is reasonable to expect that such changes will produce fluctuations in the intensity of scattered light. This idea has caused some excitement; however, experimental attempts to measure reaction rates by means of light scattering have not been successful. Apparently, in the systems studied, the changes in polarizabilities and diffusion coefficients have been quite small. The theoretical machinery for handling scattering from reacting mixtures is well developed, and the observation of reaction induced scattering may just depend on the judicious choice of systems.

As an example of polarizability fluctuation in a reaction, consider a conformational change in which form A converts into form B with the forward and reverse rates k_f and k_b, respectively, as shown below.[91]

$$A \underset{k_b}{\overset{k_f}{\rightleftharpoons}} B \tag{187}$$

The problem is similar to that encountered in Section III.D for rotational motion. Here the polarizability fluctuates between α_A and α_B because of the reaction rather than because of the rotation of an anisotropic molecule. For polarized scattering we can again use Equation 151 if the diffusion coefficients for forms A and B are similar and both conformations are optically isotropic. Suppose that P(A) is the probability that the polarizability is α_A, and $P(A|B,\tau)$ is the conditional probability that the polarizability is α_B at $t = \tau$ if it were α_A at $t = O$. The required correlation function can be written as

$$<\alpha(0)\alpha(\tau)> = \sum_{i,j = A,B} P(i)\, P(i|j,\tau)\, \alpha_i \alpha_j \qquad (188)$$

where $P(i|j,\tau)$ contains all of the information about the chemical reaction. In order to evaluate this function we assume that $\alpha(\tau)$ is a stationary Markov process. The conditional probabilities can be shown to obey the differential equation[92]

$$\frac{d\, P(i|j,\tau)}{d\tau} = \frac{-P(i|j,\tau)}{\tau_j} + \sum_k \frac{P(i|k,\tau)}{\tau_k} p_{kj} \qquad (189)$$

The mean lifetime of the kth species is τ_k; and, when a jump occurs from the kth species, p_{kj} gives the probability that it will be the jth species. In the present case $p_{AB} = p_{BA} = 1$.

The equations, which must be solved are

$$\frac{d}{d\tau} P(A|A,\tau) = -k_f P(A|A,\tau) + k_b P(A|B,\tau) \qquad (190a)$$

$$\frac{d}{d\tau} P(A|B,\tau) = -k_b P(A|B,\tau) + k_f P(A|A,\tau) \qquad (190b)$$

and a similar set with A and B reversed. In Equation 190 we have introduced the definitions $k_f = \tau_A^{-1}$ and $k_b = \tau_B^{-1}$. The standard method for solving this pair of simultaneous differential equations is to let $P(A|A,\tau) = ae^{m\tau}$ and $P(A|B,\tau) = be^{m\tau}$. Substitution of these trial functions into Equation 190 yields the roots $m_1 = O$ and $m_2 = -(k_f + k_b)$. The solutions, taking into account the initial conditions $P(A|A,O) = 1$ and $P(A|B,O) = O$, are

$$P(A|A,\tau) = P(A) + P(B)\, e^{-(k_f + k_b)\tau} \qquad (191a)$$

$$P(A|B,\tau) = P(B) \left[1 - e^{-(k_f + k_b)\tau}\right] \qquad (191b)$$

$$P(B|A,\tau) = P(A) \left[1 - e^{-(k_f + k_b)\tau}\right] \qquad (191c)$$

$$P(B|B,\tau) = P(B) + P(A)\, e^{-(k_f + k_b)\tau} \qquad (191d)$$

Substituting these functions into Equation 188 and simplifying, we obtain

$$<\alpha(0)\alpha(\tau)> = (\bar{\alpha})^2 + P(A)\,P(B)\,(\alpha_A - \alpha_B)^2\, e^{-(k_f + k_b)\tau} \tag{192}$$

where $\bar{\alpha} = P(A)\alpha_A + P(B)\alpha_B$. Now returning to Equation 151 we can obtain

$$G^{(1)}_{VV}(\tau) = AN\left[(\bar{\alpha})^2 + X_A X_B(\alpha_A - \alpha_B)^2\, e^{-(k_f + k_b)\tau}\right] e^{-D_T K^2 \tau} e^{-i\omega_0 \tau} \tag{193}$$

The mole fractions X_A and X_B are equal to the probabilities P(A) and P(B), respectively, and the equilibrium constant is

$$K_{eq} = X_B/X_A = k_f/k_b$$

The first term on the rhs of Equation 193 gives the "normal" scattering while the second term results from the chemical reaction. If $(\alpha_A - \alpha_B)^2/(\bar{\alpha})^2 << 1$, the reaction term will, of course, be difficult to detect. Notice that an extrapolation to $\theta_s = O$ should permit the rate constant to be extracted.

In transparent, i.e., nonabsorbing samples, polarizability differences such as $\alpha_A - \alpha_B$ are usually quite small. If, on the other hand, the products have significantly different absorption frequencies than the reactants, it is possible to increase the polarizability difference by tuning the exciting light into an absorption band.[6,93] This depends on the well known dispersion of the index of refraction. Consider for example the reaction of an indicator dye In^- with an H^+ ion to give the product HIn.

$$\begin{array}{ccccc} In^- & + & H^+ & \underset{k_b}{\overset{k_f}{\rightleftharpoons}} & HIn \\ \bar{C}_1 & & \bar{C}_2 & & \bar{C}_3 \end{array} \tag{194}$$

where $\bar{C}_i$ is the mean concentration of the ith species. The H^+ ion will not be detected and we assume that $D_{H^+} >> D_1 = D_3 = D_T$. The equation for $G^{(1)}_{VV}(\tau)$ in this case can be obtained either by treating the A + B ⇌ C reaction by the methods of Reference 12 or by analogy with the treatment given above. Either method gives an equation of the same form as Equation 192, but with A and B replaced by 1 and 3, respectively, and $k_f + k_b$ replaced by $\bar{C}_2 k_f + k_b$. Also, for reaction Equation 194 the mole fractions are given by

$$X_1 = \frac{k_b}{\bar{C}_2 k_f + k_b} \quad ; \quad X_3 = \frac{\bar{C}_2 k_f}{\bar{C}_2 k_f + k_b} \tag{195}$$

Because of the color change in this type of reaction it may be possible to choose a wavelength that will permit the reaction term to be measured.

It has been suggested that changes in diffusion coefficients of more than 30% may be found in macromolecular dimerization and isomerization reactions, and that this could lead to detectable changes in quasi-elastic light scattering.[94] An even more likely possibility is that the reacting molecules will have different electrophoretic mobilities and the ELS spectrum will depend on the reaction rate.[11] Calculated curves for different reaction rates have been presented, but as yet no experimental examples have been reported. A formulation of the problem which includes all of these effects is discussed in detail in Chapter 6 of *Dynamic Light Scattering.*[12]

H. Experimental Capabilities and Limitations

1. Light Sources and Detectors

Collimated beams of continuous wave (CW), monochromatic radiation are required in QLS, and lasers are the only practical sources of such radiation. Contrary to common belief, conventional sources could be used in light-beating spectroscopy (LBS) if sufficient intensity were available.[95-97] Restrictions on the bandwidths of the sources are, in fact, not very severe. For example, dye lasers with tuning elements but without high-resolution etalons are satisfactory sources even though the bandwidths are of the order of 10^{10} Hz.[98] The bandwidth of the source would clearly dominate in optical spectroscopy, but in intensity fluctuation spectroscopy (IFS) extremely high resolution can, in fact, be obtained with fairly broadband sources.

A simple analysis of light scattering using a quasi-monochromatic source reveals the requirements and limitations. The electric field of the laser light at the sample can be written as

$$E_L(t) = E_L^0 \, e^{-i[\omega_0 t + \phi_L(t)]} \tag{196}$$

where $\phi_L(t)$ is a time-dependent phase fluctuation, which is characteristic of instabilities in the source. It is this phase fluctuation which is responsible for most of the optical line width. In IFS $\phi_L(t)$ is of no consequence since the measured quantity is the intensity $I = E^*E$. The important factor turns out to be the amplitude fluctuation in the source, i.e., fluctuations in $|E_L^\circ|$. The electric field of the scattered light can be expressed as

$$E_S(t) = E_L^0(t) \, e^{-i[\omega_0 t + \phi_L(t)]} \, f_M(t) \tag{197}$$

where $f_M(t)$ describes the modulation resulting from the scattering medium. If fluctuations in the source and the scattering medium are independent and Gaussian distributed, the reduced second-order correlation function can be factored as follows:

$$\begin{aligned} g_S^{(2)}(\tau) &= g_L^{(2)}(\tau)\, g_M^{(2)}(\tau) \\ &= \left[1 + |g_L^{(1)}(\tau)|^2\right]\left[1 + |g_M^{(1)}(\tau)|^2\right] \end{aligned} \tag{198}$$

Now suppose that $|g_L^{(1)}(\tau)| \sim e^{-\tau/\tau_L}$ and $|g_M^{(1)}(\tau)| \sim e^{-\tau/\tau_M}$. If only low frequency amplitude fluctuations are present in the laser, then $\tau_L >> \tau_M$, and we find that $g_S^{(2)}(\tau) \simeq 2$

$g_M^{(2)}(\tau)$. At the opposite limit high frequency amplitude fluctuations give $\tau_L << \tau_M$ and $g_S^{(2)}(\tau) \simeq g_M^{(2)}(\tau)$. Therefore, in principle $g_M^{(2)}(\tau)$ can be obtained using either narrow or broadband sources. In general when laser intensity fluctuations are present the scattered light is not Gaussian.

In practice two difficulties arise. First, the intensity fluctuations in the laser may have frequency components in the same range as the fluctuations of interest in the sample, i.e., τ_L may be roughly equal to τ_M. Usually source fluctuations of up to 2% root-mean-square (rms) can be tolerated. Some lasers are inherently noisy while others have noise which results from malfunctions in their power supplies. These fluctuations are unimportant in many applications but are catastrophic in IFS. This kind of problem must be solved at the source or at least before the beam reaches the sample. It is an important consideration in selecting a laser system, but one should be aware that the noise level for a given laser may change with time. An effective but expensive remedy is to insert an electro-optic noise reduction system, i.e., a "noise eater", in the optical path. For example, the Coherent Associates Model 307 provides a 30 to 40 dB reduction in noise below 100 kHz.

The second problem limits the laser bandwidths which can be used. Recall that in nonideal situations the Siegert relation is given by

$$g^{(2)}(\tau) = 1 + \beta \, | g^{(1)}(\tau) |^2 \qquad (105')$$

where β can become very small for samples which are not coherently illuminated. The effective scattering volume depends on the coherence length $\ell_c = c/(n\,\Delta\nu)$ of the incident light where n is the index of refraction of the medium and $\Delta\nu$ is the laser band width. In order to have a coherence length of 1 mm the laser bandwidth must be less than 8 cm^{-1}. Of course, a narrow band filter can be inserted if necessary, but this may reduce the useable intensity to an unsatisfactory level.

In Table 6 we list some commercially available CW lasers which may be useful in IFS. Most experimental studies to date have made use of He-Ne lasers, which are quiet and reliable though limited in power. Interest in the other lasers arises because other wavelengths are sometimes required in order to avoid absorption or to search for resonance enhancement. Dye lasers have the advantage of tunability, but with a given dye and the corresponding set of mirrors the range is somewhat limited. Also, unacceptable intensity fluctuations often result from bubbles in the dye jet and other mechanical instabilities, especially when operating near threshold.

The incident beam in a light scattering experiment is expected to be quasi-monochromatic, well-defined in space, and polarized in a known direction. Lasers are less than perfect in these characteristics. For example, gas lasers emit light at frequencies corresponding to plasma oscillations as well as at the primary beam frequency ω_0, and they often have secondary output beams in various directions. Dye lasers always produce some background fluorescence. Therefore, it is advisable to prepare the beam by using narrow band filters, apertures, polarization rotators, and polarizing prisms. A vibration isolation table also may be required if there is evidence of effects of mechanical oscillations in the power spectra or correlation functions.

The light detectors are essentially always photomultiplier tubes (PMT). These are photo emissive devices coupled with electron multipliers. When a photon strikes the photocathode, an electron is emitted with a quantum efficiency of as high as 0.3. The multiplier sections typically have gains of about 10^6 so that a measurable current is produced at the anode. In a photon counting system the pulses resulting from single photons are selected, amplified, and counted. In analog systems the average anode

Table 6
CONTINUOUS LASERS THAT MAY BE SUITABLE FOR INTENSITY FLUCTUATION SPECTROSCOPY

Laser	Wavelength range (nm)	Comments
Argon ion (gas)	454.5-528.7	Discrete lines with maximum power at 488 and 514.5;
	351.1-363.8 } 1090	requires special optics.
Krypton ion* (gas)	476.2-799.3	Discrete lines with maximum power at 647.1;
	350.7 } 364.4	requires special optics.
Helium-Cadmium*	441.6	
(gas)	325	Requires special optics.
Helium-Neon (gas)	632.8	Power <75 mW but stable and reliable;
	611.8; 640.1 } 1084, 1152	requires special optics.
Dye laser (liquid)	430-900	Requires about ten different dyes and several sets of mirrors to cover this range; also, intense pump lasers in UV and visible (both green and red) are required.
Semiconductor diode laser*	820-870	Single line somewhere in this region; a linewidth of ~0.1 nm at 15 mW can be obtained.
Neodymium Yag*	1060	

* Use in intensity fluctuation experiments has not been reported.

current i(t) is measured. Spectrum analyzers usually operate on i(t) to determine the power spectrum $I^{(2)}(\omega)$. In this case the requirements on the PMT are not very severe, and inexpensive tubes such as the RCA 7265® can be used. With autocorrelators however, there are advantages to using photon counting electronics, and the response time of the PMT becomes important. Favorite tubes for photon counting applications have been the ITT FW130® and the Channeltron BX-7500®. Both of these tubes have S20 response and, therefore, have adequate quantum efficiency at 632.8 nm; but their sensitivities decrease rapidly with increasing wavelength. The newer RCA C31034(A)® has a GaAs photocathode that permits a spectral response range of 200 to 950 nm. In addition, it is superior in quantum counting efficiency and dead time.[99] This appears to be the tube of choice in spite of its relatively high price and the requirement that it be cooled to −20°C to reduce the dark count. At wavelengths longer than 950 nm one of the Varian InGaAsP® tubes such as the VPM-159A2 would probably be the best detector for photon counting applications.

2. Spectrum Analyzers and Correlators

Spectrum analyzers are commonly used in laser velocimetry where frequency shifts appear and several frequency components may be present, e.g., in ELS. The simplest spectrum analyzer consists of a narrow band amplifier, the center frequency of which can be scanned across the frequency range of interest to obtain $I^{(2)}(\omega)$. This was the basis of the old HP-302A wave analyzers which used an electric motor to scan the amplifier at rates up to 1 kHz/min. The same idea was implemented in the fast scanning cathode ray tube (CRT) type spectrum analyzers such as the Singer Panaramic MF-5®. Much greater efficiency and resolution are now available in "time compression" type spectrum analyzers which record the incoming signal and then play it back at different rates to pick out the various frequency components. An example of this type of analyzer is the 400 channel Honeywell SAICOR® which provides 11 frequency ranges from 0 to 20 Hz to 0 to 1 MHz and has built-in signal averaging capability. For very high resolution in the frequency range below 50 kHz the realtime FFT spectrum analyzers offer the best performance. These microprocessor based units perform a fast Fourier transformation on an input record containing n points to produce a power spectrum having n/2 points. Examples of FFT analyzers are the PAR 4512 and 4513 and the recently announced HP 3582 A.

In experiments where the correlation function $G^{(2)}(\tau)$ is a monotonically decreasing function of time it is most efficient to determine this function directly from the PMT output. Digital correlators calculate approximations to $G^{(2)}(\tau)$ as discussed in Section III.A.3. In photocurrent autocorrelation an analog to digital (A/D) converter is required in the correlator while with photon counting systems the intensity $I_s(t)$ of the incident light is proportional to the number of counts $n(t,\Delta T)$ in the time interval ΔT at t, and the autocorrelation function is given by $G^{(2)}(\tau) = \langle n(0,\Delta T)\, n(\tau,\Delta T)\rangle$ except for a proportionality constant. A correlator calculates and stores in the kth channel the function

$$C(k\Delta T) = \sum_{i=1}^{N_s} n(t_i,\Delta T)\, n(t_i + \tau,\Delta T) \tag{199}$$

where $\tau = k\Delta T$ and $n(t_i,\Delta T)$ is the number of photocounts measured in the time interval ΔT starting at time t_i. The Siegert relation is usually assumed and the equation

$$C(\tau) = A[1 + \beta\, |g^{(1)}(\tau)|^2] \tag{200}$$

must be used to obtain $|g^{(1)}(\tau)|$. In photocurrent autocorrelation both A and β must be determined from a least squares fit unless special filtering has been introduced to remove the DC component. One of the advantages of photocount autocorrelation is that A can be calculated directly. This is possible since the correlator keeps up with the total number of counts.

$$N_T = \sum_{i=1}^{N_s} n(t_i,\Delta T) \tag{201}$$

and the number of terms N_s in the summation. Therefore, $\langle n\rangle = N_T/N_s$, $C(\tau) = N_s G^{(2)}(\tau)$, and Equation 200 can be written as

$$N_S G^{(2)}(\tau) = C(k\Delta T) = \frac{N_T^2}{N_S}[1 + \beta\, |g^{(1)}(\tau)|^2] \tag{202}$$

High speed digital correlators are now available which permit sample times ΔT as short as 100 nsec. The units manufactured by Langley-Ford and by Malvern are real time in the sense that processing occurs simultaneously for all of the channels, and the processing efficiency does not change significantly when the shorter time intervals are selected. A typical number of channels is 64 and expansion is offered as an option. A very important feature is that the last few channels can be shifted out to greater time delays. This permits A in Equation 200 to be determined independently of the calculation described above. If the two values differ, contamination of the sample or intensity fluctuations in the laser should be considered.

Thus far we have assumed that the full correlation function $C(k\Delta T)$ is calculated by the correlator. In practice this means that 4-bit shift registers are used and $n(t,\Delta T)$ is stored as an integer from 0 to 15. If more than 15 counts occur in the interval ΔT, a correction is required. This scheme is called 4×4 or full processing. Furthermore, some correlators gain speed by operating in the single or double clipped mode. This means that $n(t,\Delta T)$ is stored as 1 or 0 depending on whether its magnitude exceeds some predetermined level, i.e., the "clipped" count rate is defined as

$$n_k(t,\Delta T) = 1 \text{ for } n(t,\Delta T) > k$$

$$= 0 \text{ for } n(t,\Delta T) \leq k$$

The single and double clipped correlation functions are defined by Equations 203 and 204, respectively, where ΔT has been suppressed.

$$g_k^{(2)}(\tau) = \langle n_k(t)\, n(t+\tau)\rangle / \langle n_k\rangle \langle n\rangle \tag{203}$$

$$g_{kk'}^{(2)}(\tau) = \langle n_k(t)\; n_{k'}(t+\tau)\rangle / \langle n_k\rangle \langle n_{k'}\rangle \tag{204}$$

If $\langle n\rangle \ll 1$, both $g_k^{(2)}(\tau)$ and $g_{kk'}^{(2)}(\tau)$ are equivalent to $g^{(2)}(\tau)$, but as $\langle n\rangle$ increases distortions can occur. However, in the special case of Gaussian light it can be shown that

$$g_k^{(2)}(\tau) = 1 + \frac{(1+k)}{(1+\langle n\rangle)} \, | g^{(1)}(\tau) |^2 \tag{205}$$

The time dependence is not distorted and for $k = \langle n\rangle$, $g_k^{(2)}(\tau)$ is identical to $g^{(2)}(\tau)$. At higher count rates the double clipped function will always introduce some distortion.

3. Special Requirements

a. Coherence Areas

A major difference between the experimental requirements for classical light scattering and intensity (interference) fluctuation spectroscopy (IFS) is the importance of coherence areas in the latter. The following facts characterize the IFS experiment:

1. Light scattered from a coherently illuminated volume gives a characteristic speckle pattern, i.e., an array of bright nonoverlapping spots.
2. The temporal fluctuations of interest appear in one speckle, i.e., in one coherence area, and the signal-to-noise ratio (S/N) does not increase when more than one speckle is detected.

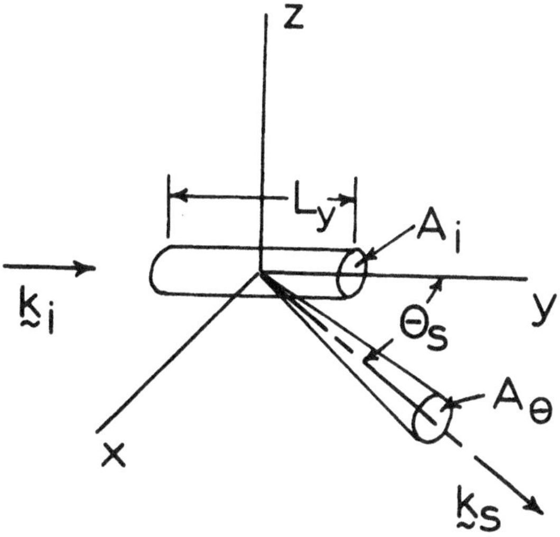

FIGURE 21. The approximate shape of an illuminated volume having the cross-sectional area A_i and the length L_y.

If a detector collects light in the solid angle $\Delta\Omega$ at the distance R, the area of the detector is given by $A_\theta = R^2 \Delta\Omega$. The coherence area and the coherence angle are denoted by A_θ and $\Delta\Omega_{coh}$, respectively, and the number of coherence areas collected is $N_{coh} = A_\theta/A_{coh}$. Since S/N does not increase as N_{coh} increases for $N_{coh} > 1$, and may in fact decrease, the important quantity is the amount of power scattered into one coherence area. From Equation 16 we see that the power P_θ in the solid angle $\Delta\Omega$ is given by

$$P_\theta = \mathcal{R}_\theta L_y P_i \Delta\Omega \tag{206}$$

where L_y is the length of the scattering volume. The Rayleigh ratio $\mathcal{R}_\theta$ is characteristic of the sample, but the other quantities are, to a certain extent, under our control.

Consider the geometry shown in Figure 21. The illuminated volume has the dimensions L_x, L_y, and L_z and the cross-sectional area $A_i \sim L_x L_z$ so that the magnitude of the volume is given by $V = A_i L_y$. The calculation of the coherence angle is not trivial, but a detailed treatment has been given by Lastovka.[100] A summary can also be found in Chu's book.[13] The basic idea is that the speckle pattern is an instantaneous Fourier expansion of the distribution of scatterers in the illuminated volume. Only discrete values of $\underset{\sim}{K}$ are permitted in this expansion, and these correspond to the speckles. The speckles turn out to have finite size since the Fourier integrals span a volume which is finite in extent. The complete analysis gives

$$\Delta\Omega_{coh} = \frac{\lambda^2}{L_y L_z[\sin\theta_s + (L_x/L_y)\cos\theta_s]} \tag{207}$$

It is interesting and significant that the denominator in Equation 207 is equal to R^2

times the solid angle Ω subtended by the illuminated volume at the detector, i.e., $A_{coh} = \lambda^2/\Omega$. When Equations 206 and 207 are combined, we obtain an expression for the power scattered into one coherence area.

$$P_\theta(\Delta\Omega_{coh}) = \frac{\mathcal{R}_\theta P_i \lambda^2}{L_z[\sin\theta_s + (L_x/L_y)\cos\theta_s]} \tag{208}$$

Equation 208 suggests that L_x and L_z should be minimized, but that L_y can be increased without penalty. Also, the power in $\Delta\Omega_{coh}$ is seen to increase as θ_s decreases. It turns out that the wavelength dependence of $P_\theta(\Delta\Omega_{coh})$ is much less than expected since the λ^2 in $\Delta\Omega_{coh}$ partially cancels the λ^{-4} which appears in $\mathcal{R}_\theta$. In fact, a complete analysis, taking into account the wavelength dependence of the characteristic decay time of $G^{(2)}(\tau)$ and the energy per photon hc/λ, shows that the statistical accuracy in a photon correlation spectroscopy (PCS) experiment of fixed duration is independent of the wavelength.[37] This, of course, assumes that the quantum efficiency of the PMT is constant.

The precise dependence of S/N on N_{coh} in a real experiment is not known. Experience indicates that it is best to collect a few coherence areas, e.g., somewhere in the range 2 to 5. It is clear that large values of N_{coh} are not beneficial.

b. Stray Light

Uncontrolled light can mix with the light from the scattering volume to give a heterodyne component. This may, of course, have a large effect on the measured line width. Also, back scattered light from the exit wall of the sample cell can have sufficient intensity to give secondary scattering at the angle $\theta'_s = \pi - \theta_s$. These effects are sufficient to indicate that a well-characterized scattering experiment requires that stray light be eliminated as much as possible. Since each air-glass interface can scatter about 4% of the incident light, it is important to keep these interfaces far from the scattering volume. A cell holder designed for this purpose has been described by Jolly and Eisenberg.[101] The important feature is that their cylindrical sample cell (10 mm or 29 mm diameter) is suspended in an index of refraction matching fluid which is contained in a cylindrical tank having a diameter of 125 mm. The incident beam is defined by an aperture in the bath, and after the sample cell there is a light trap, also in the bath.

Another solution to the problem of stray light, especially in connection with low angle scattering, has been discussed by Kaye et al., and incorporated into the Chromatix KMX-6® photometer.[17,102] Their idea is to contain the sample inside a spacer between two thick fused-silica windows. The exiting light on axis (θ_s = O) is trapped while the light scattered into the cone between θ_s and $\theta_s + \Delta\theta_s$ is collected through a defining annulus. It is reported that scattering angles from 2° to 7° can be covered and that the sample volume is only 150 μl. With this type of optical arrangement the angular factor $(1 + \cos^2\theta_s)/2$ must be included in the expression for the scattered intensity regardless of the polarization state of the incident light (see Figure 17). Also, the effect of the number of coherence areas collected must be considered as previously discussed.

c. Particulate Contamination

Dust and other large particles can be a severe problem in QLS since the scattering intensity depends on the square of the mass of the scattering particle. Closed loop recirculation systems containing millipore or other types of filters are fairly effective in removing dust, but these systems must be rigorously sealed. The performance of such a system can be monitored by observing the illuminated volume with a small microscope. Another convenient test is to compare the scattered intensities at $\theta_s = 45°$ and 135° using polarized incident light since for small particles the intensity should be independent of the scattering angle. If the sample of interest is known to be monodisperse, the variance $\mu_2/<\Gamma>^2$, obtained from an analysis of the measured correlation function, can also be used as a criterion of sample quality. Still another effective test is to compare the normalization constant A in Equation 200 obtained from the baseline at large τ with the calculated value N_T^2/N_s. If large, slowly moving particles are present, the correlation function will contain slowly decaying components which may contribute to $G^{(2)}(\tau)$ at large but not infinite τ; and effectively establish a baseline at a higher level than that calculated by Equation 202. In such cases the measured baseline should be used; however, the only good solution to the dust problem is to make sure that the dust is removed.

d. Absorbing Samples

When a sample has significant absorption at the wavelength of the incident light, the efficiency of the PCS experiment is greatly reduced and effects appear which may complicate the interpretation of the results. The most obvious problem is the loss of signal, which results from the attenuation of both the incident and the scattered light. If the scattering volume is at the center of a cell having a diameter of 1.0 cm, an absorption coefficient (absorbance/cm) of 1.0 reduces the signal by a factor of 10. It is not permissible to compensate for this effect by increasing the incident power since this causes heating of the sample and leads to additional problems. The best approach is to minimize the path lengths for the incident and scattered light by judicious cell design. A fairly satisfactory arrangement is to use a cylindrical jacketed cell, e.g., the Hellma cell no. 165, and to pass the focused laser beam parallel to the axis of the cell but only ~ 1 mm from the wall. The scattered light can then be collected through the side of the water jacketed at fairly large scattering angles or at smaller angles through the flat front face of the water jacket. The total path length in the sample can be held to about 2.5 mm in this way.

Heating effects can be serious at even moderate power levels. The diffusion coefficient for an aqueous solution increases roughly 3% per degree centrigrade, so the temperature rise alone is significant. It is not easy to calculate the temperature rise in the scattering volume for a given power level, but an estimate can be based in the equations of Gordon et al.[103-105] For a typical experiment involving an aqueous solution with an absorption coefficient of unity and a laser beam having a diameter of 0.1 mm, the temperature rise turns out to be 0.8°C at 2 mm from the entrance face for an incident power of 10 mW. This crude estimate only serves to emphasize that power levels should be kept low, and that measured diffusion coefficients should be extrapolated to zero power.[117]

For aqueous samples the temperature rise is usually the major effect; however, convection and thermal lensing must also be considered. The increased temperature in the laser beam results in reduced density which in turn gives an upward buoyant force. As discussed in Section III.C.1, uniform velocity cannot be detected in homodyne experiments; however, convection may produce a distribution of velocities. Again, low incident power is recommended. Also, because of viscous forces the use of a tightly focused laser beam will minimize velocity gradients in the scattering volume. In any event it is important to keep the scattering plane horizontal so that the dot product $\underset{\sim}{K} \cdot \underset{\sim}{v}_d$ will be zero. This removes the effect of the vertical component of the convection velocity in both homodyne and heterodyne experiments, even when a distribution of velocities is present.

Thermal lensing results from gradients in the index of refraction which exist because of the steady-state temperature distribution. Since the index is usually lowest along the axis of the laser beam, the sample behaves as a diverging lens. The result is that the beam may increase in diameter as it passes through the sample and a distribution of scattering angles may contribute to the light scattered in a given direction. At higher power levels severe divergence and even aberrations may occur.

APPENDIXES

A. The Polarizability Tensor

The need for tensors arises when the induced dipole is not in the same direction as the electric field. The tensor $\underset{\approx}{\alpha}$ permits both the magnitude and the direction of the induced dipole to be calculated when the direction and magnitude of the electric field are specified. The equation $\underset{\sim}{p} = \underset{\approx}{\alpha} \cdot \underset{\sim}{E}$ is shorthand notation for the set of three equations:

$$\begin{aligned} P_x &= \alpha_{xx} E_x + \alpha_{xy} E_y + \alpha_{xz} E_z \\ P_y &= \alpha_{yx} E_x + \alpha_{yy} E_y + \alpha_{yz} E_z \\ P_z &= \alpha_{zx} E_x + \alpha_{zy} E_y + \alpha_{zz} E_z \end{aligned} \tag{209}$$

where the nine quantities α_{ij} (i,j = x,y,z) are elements of the polarizability tensor. It is common to express Equation 209 in matrix form as:

$$\begin{pmatrix} P_x \\ P_y \\ P_z \end{pmatrix} = \begin{pmatrix} \alpha_{xx} & \alpha_{xy} & \alpha_{xz} \\ \alpha_{yx} & \alpha_{yy} & \alpha_{yz} \\ \alpha_{zx} & \alpha_{zy} & \alpha_{zz} \end{pmatrix} \cdot \begin{pmatrix} E_x \\ E_y \\ E_z \end{pmatrix} \tag{210}$$

In transparent media these tensors are usually real and symmetric, i.e., $\alpha_{ij} = \alpha_{ji}$ and $\alpha_{ij} = \alpha_{ij}^*$.

The form of a tensor depends on the coordinate system chosen for its representation. Since both E and p are specified in the laboratory fixed coordinate system, we chose xyz in Equations 209 and 210 as axes in this frame. Suppose, for example, that the scatterer has $\alpha_{xy} = \alpha_{xz} = O$. If the incident light is polarized in the z-direction, as shown in Figure 1, the induced dipole will be given by $p = (\alpha_{yz}\,\hat{j} + \alpha_{zz}\,\hat{k})\,E_z$. It is always possible to choose a coordinate system so that the matrix $[\alpha]$ has the diagonal form, i.e.,

$$[\alpha] = \begin{pmatrix} \alpha_{XX} & 0 & 0 \\ 0 & \alpha_{YY} & 0 \\ 0 & 0 & \alpha_{ZZ} \end{pmatrix} \tag{211}$$

Here XYZ, which are called the principal axes for the polarizability tensor, have the special property that a field directed along one of the axes will induce a dipole along the same axis (see Figure 18).

The principal axes for a molecule can usually be determined without difficulty by considering molecular symmetry, e.g., the six-fold rotational axis of benzene is one of the principal axes of the polarizability tensor. The elements of the polarizability tensor in the laboratory frame can then be related to the elements in the molecule fixed frame by means of the equation:[106]

$$\alpha_{ij} = \sum_{i',j'} \alpha_{i'j'} \ell_{i'i} \ell_{j'j} \qquad (212)$$

where $\ell_{i'i}$ is the cosine of the angle between i′ and i axes. However, regardless of the coordinate system chosen, certain properties of the matrix $[\alpha]$ are unchanged. The invariants of interest here are the mean polarizability α defined by[107]

$$\alpha = \frac{1}{3}(\alpha_{xx} + \alpha_{yy} + \alpha_{zz}) \qquad (213)$$

and the anisotropy β defined by

$$\beta^2 = \frac{1}{2}[(\alpha_{xx} - \alpha_{yy})^2 + (\alpha_{yy} - \alpha_{zz})^2 + (\alpha_{zz} - \alpha_{xx})^2 + 6(\alpha^2_{xy} + \alpha^2_{yz} + \alpha^2_{zx})] \qquad (214)$$

In liquids and gases the situation is somewhat more complicated since the molecules are randomly oriented with respect to the laboratory coordinate system. Any measurement on a bulk system which depends on combinations of components of the polarizability tensor must in fact determine only averages of the combination over all orientations. The intensity of the scattered radiation is an important example. According to Equation 9, the intensity of light scattered by an isotropic scatterer depends on α^2. When anisotropy is present, the scattered light contains both polarized and depolarized components. Consider the geometry shown in Figure 1. Light polarized in the z-direction is said to be vertically polarized and is denoted by the index V. Similarly, light polarized in the x-direction is horizontally polarized and is denoted by H. For simplicity let $\chi = \theta_s = \pi/2$. If the incident light is vertically polarized, the intensity of the scattered light which is polarized vertically is proportional to $\overline{\alpha^2}_{zz}$ where the subscripts refer to the laboratory coordinate system and the bar indicates an average over molecular orientations. Similarly, the intensity of the scattered light which is polarized horizontally is proportional to $\overline{\alpha^2}_{zy}$. When vertically polarized incident light is used, the depolarization ratio is defined as

$$\rho_V = \frac{I_{VH}}{I_{VV}} = \frac{\overline{\alpha^2_{zy}}}{\overline{\alpha^2_{zz}}} \qquad (215)$$

When naturally polarized incident light is used, the depolarization ratio is given by

$$\rho_n = \frac{I_{VH} + I_{HH}}{I_{VV} + I_{HV}} = \frac{\overline{\alpha_{zy}^2} + \overline{\alpha_{xy}^2}}{\overline{\alpha_{zz}^2} + \overline{\alpha_{xz}^2}} \qquad (216)$$

When two indexes appear on the scattered intensity, the first indicates the direction of polarization of the incident light and the second refers to the scattered light.

The problem of calculating averages for products of tensor components for rotating molecules arises in many branches of physical chemistry, and the solution has been given by many authors. A complete derivation can be found in Reference 106. The averages required in light scattering are the following:[107]

$$\overline{\alpha_{xx}^2} = \overline{\alpha_{yy}^2} = \overline{\alpha_{zz}^2} = \alpha^2 + \frac{4}{45}\beta^2 \qquad (217)$$

$$\overline{\alpha_{xy}^2} = \overline{\alpha_{yz}^2} = \overline{\alpha_{zx}^2} = \frac{\beta^2}{15} \qquad (218)$$

$$\overline{\alpha_{xx}\,\alpha_{yy}} = \overline{\alpha_{yy}\,\alpha_{zz}} = \overline{\alpha_{zz}\,\alpha_{xx}} = \alpha^2 - \frac{2}{45}\beta^2 \qquad (219)$$

Therefore, depolarization ratios can be expressed in terms of the invariants α and β, and the intensity of the depolarized component of the scattered light can be used to determine β. Using Equations 217 and 218, depolarization ratios can be written as:

$$\rho_V = \frac{3\beta^2}{45\alpha^2 + 4\beta^2} \qquad (220)$$

$$\rho_n = \frac{6\beta^2}{45\alpha^2 + 7\beta^2} \qquad (221)$$

In many scattering experiments the incident light is vertically polarized, but the scattered light at $\theta_s = \pi/2$ is collected without the use of a polarization analyzer. If the anisotropy $\beta \neq 0$, the total scattered intensity is greater than that resulting from the mean polarizability. It is sometimes necessary to derive the Rayleigh ratio expected for isotropic scatterers having the polarizability α from the measured total intensity. This can be done as follows. For the total intensity we have

$$I_{TOTAL} = I_{VV} + I_{VH} \propto \overline{\alpha_{zz}^2} + \overline{\alpha_{zy}^2} = \frac{45\alpha^2 + 7\beta^2}{45}$$

Since the Rayleigh ratio is defined by $\mathcal{R}_\theta = \bar{I}_s R^2/\bar{I}_0$, we can write

$$\frac{\mathcal{R}_\theta\ (\beta \neq 0)}{\mathcal{R}_\theta\ (\beta = 0)} = \frac{45\ \alpha^2 + 7\ \beta^2}{45\ \alpha^2} = \frac{1 + \rho_v}{1 - \frac{4}{3}\rho_v} \quad (222)$$

This ratio, which is well known in classical light scattering, is known as the Cabannes factor. When naturally polarized incident light is used, the Cabannes factor has the form[108]

$$\left[\frac{6 + 6\ \rho_n}{6 - 7\ \rho_n}\right]$$

Thus, the Rayleigh ratio $\mathcal{R}_\theta(\beta = 0)$ for scattering by concentration fluctuations alone can easily be derived from the experimental Rayleigh ratio by using Equation 222 if the depolarization ratio is known.

B. Electromagnetic Waves

The purpose of this appendix is to show how the susceptibility χ, the dielectric constant k_e, the polarizability α, and the index of refraction n are related.[109] For simplicity we consider a plane wave propagating in the +y-direction with the velocity v. The evolution of the amplitude U(y,t) for this wave obeys the equation:

$$\frac{\partial^2 U(y,t)}{\partial y^2} = \frac{1}{v^2}\frac{\partial^2 U(y,t)}{\partial t^2} \quad (223)$$

For light propagating in the +y-direction in a vacuum Maxwell's equations show that the amplitude of the electric field is described by

$$\frac{\partial^2 E(y,t)}{\partial y^2} = \mu_0 \epsilon_0 \frac{\partial^2 E(y,t)}{\partial t^2} \quad (224)$$

where $\mu_0 = 4\pi \times 10^{-7}$ H/m is the permeability of the vacuum and $\varepsilon_0 = 8.854 \times 10^{-12}$ F/m is the permittivity of the vacuum. A comparison of Equations 223 and 224 indicates that the speed of light in vacuum is given by

$$c = \frac{1}{\sqrt{\epsilon_0 \mu_0}} = 2.997 \times 10^8 \text{ m/s} \quad (225)$$

The solution of Equation 224 has the form

$$E(y,t) = E_0 \cos(k_0 y - \omega_0 t) = E_0 \operatorname{Re}\left[e^{i(k_0 y - \omega_0 t)}\right] \quad (226)$$

where $k_o = 2\pi/\lambda_o$, λ_o is the wavelength in vacuum, and Re means "the real part of". Substitution of Equation 226 into 224 immediately gives $k_o = \omega_o/c$ as expected.

In nonconducting media, Equation 224 takes the form:

$$\frac{\partial^2 E(y,t)}{\partial y^2} = \mu_o \frac{\partial^2}{\partial t^2} (\epsilon_o E + P) \tag{227}$$

The quantity $\varepsilon_o E + P$ is known as the electric displacement and is assigned the symbol D. The displacement is also written as

$$D = \epsilon E = \epsilon_o E + P \tag{228}$$

where ε is the permittivity of the medium. Assuming that E is sufficiently small that the response of the medium is linear, the polarization P is proportional to E and can be written as:

$$P = \epsilon_o \chi E = N \alpha E \tag{229}$$

Equation 229 serves to define the susceptibility χ per unit volume and the polarizability α per scatterer. N is the number of scatterers per unit volume. At this point we combine Equations 228 and 229 to define the dielectric constant k_e, which is also called the relative permittivity.

$$k_e = \frac{\epsilon}{\epsilon_o} = (1 + \chi) = \left(1 + \frac{N\alpha}{\epsilon_o}\right) \tag{230}$$

With these definitions Equation 227 can be written as

$$\begin{aligned} \frac{\partial^2 E(y,t)}{\partial y^2} &= \mu_o \epsilon_o (1 + \chi) \frac{\partial^2 E(y,t)}{\partial t^2} \\ &= \frac{k_e}{c^2} \frac{\partial^2 E(y,t)}{\partial t^2} \end{aligned} \tag{231}$$

By considering only the electric polarization of the medium we have already made the assumption that the permeability μ does not differ from its value μ_o in vacuum. This is a reasonable assumption for nonmagnetic materials.

The solution of Equation 231 has the same form as that for Equation 224 except that we now must allow for both retardation and absorption of the incident light. These features can be incorporated by permitting k_e and hence χ to have both real and imaginary parts. With this change the solution still has the form shown in Equation 226, and the substitution of 226 into 231 gives $k^2c^2/\omega^2_o = k_e$. It is consistent with common usage to define the index of refraction n and the extinction coefficient $\varkappa$ through the equation

$$kc/\omega_0 = n + i\kappa \tag{232}$$

Then the wave amplitude in Equation 226 becomes

$$\begin{aligned} E(y,t) &= E_0 \,\mathrm{Re}\left\{ e^{i\left[\frac{\omega_0}{c}(n + i\kappa)y - \omega_0 t\right]} \right\} \\ &= E_0\, e^{-\omega_0 \kappa y/c}\, \mathrm{Re}\left[e^{i(nk_0 y - \omega_0 t)} \right] \end{aligned} \tag{233}$$

The intensity I, which is proportional to E^2, is seen to decay as $e^{-2\omega_0\kappa y/c}$; and the wavelength λ is given by $\lambda = \lambda_0/n$ or $k = nk_0$. Notice also that Equation 231 indicates that the velocity v for light in the medium is related to the speed of light c in vacuum by $k_e/c^2 = 1/v^2$ or $c/v = \sqrt{k_e}$. By definition c/v is equal to n, and therefore $n^2 = k_e$ in the absence of absorption.

For macromolecules in solution it is convenient to write

$$\begin{aligned} k_e &= (\chi_{solvent} + \chi_{solute}) + 1 \\ &= (\chi_{solvent} + 1) + \chi_{solute} \end{aligned} \tag{234}$$

Then

$$k_e - (k_e)_{solvent} = \chi_{solute}$$

$$n^2 - n_0^2 = N\alpha/\epsilon_0$$

where N is again the number of scatterers per unit volume.

C. Thermodynamic Relations

1. The Relationship of $(\partial^2 A/\partial C^2)_{T,V}$ to $(\partial \mu_1/\partial C)_{T,V}$

Since V is constant, the number of moles of components 1 and 2 are related by

$$\delta V = n_1 \bar{V}_1 + n_2 \bar{V}_2 \tag{235}$$

where $\bar{V}_1$ and $\bar{V}_2$ are the partial molar volumes of components 1 and 2, respectively. By convention the solvent is denoted by 1 and the solute by 2. Therefore, a change in the concentration of one species will be reflected in a change in the other according to

$$dn_1 = -\frac{\bar{V}_2}{\bar{V}_1}\, dn_2 \tag{236}$$

The change in Helmholtz free energy associated with the composition change at constant volume and temperature is given by

$$dA = \mu_1 dn_1 + \mu_2 dn_2 \qquad (237)$$

where μ_1 and μ_2 are the chemical potentials of the solvent and solute, respectively. By combining Equations 236 and 237 we obtain

$$dA = \left[\mu_2 - \frac{\bar{V}_2}{\bar{V}_1}\mu_1\right] dn_2 \qquad (238)$$

It is also apparent that dn_2 can be expressed in terms of the concentration C since $(n_2/\delta V) = C/M$ and $dn_2 = (\delta V/M)dC$. Substituting for dn_2 in Equation 238 gives

$$\left(\frac{\partial A}{\partial C}\right)_{T,V} = \left[\mu_2 - \frac{\bar{V}_2}{\bar{V}_1}\mu_1\right]\frac{\delta V}{M} \qquad (239)$$

Differentiation of Equation 239 with respect to C gives

$$\left(\frac{\partial^2 A}{\partial C^2}\right)_{T,V} = \left[\left(\frac{\partial \mu_2}{\partial C}\right)_{T,V} - \frac{\bar{V}_2}{\bar{V}_1}\left(\frac{\partial \mu_1}{\partial C}\right)_{T,V}\right]\frac{\delta V}{M} \qquad (240)$$

It should also be noted that the partial molar volumes depend on the concentration, but the fluctuations are too small for this to be important. We also note that the differentials of chemical potentials are related by the Gibbs-Duhem equation

$$n_1 d\mu_1 + n_2 d\mu_2 = 0 \qquad (241)$$

so $d\mu_2 = -n_1/n_2\, d\mu_1$. Substitution for $d\mu_2$ in Equation 240 gives

$$\left(\frac{\partial^2 A}{\partial C^2}\right)_{T,V} = -\frac{\delta V}{M}\left[\frac{n_1\bar{V}_1 + n_2\bar{V}_2}{n_2\bar{V}_1}\right]\left(\frac{\partial \mu_1}{\partial C}\right)_{T,V} \qquad (242)$$

and recalling that $C = Mn_2/(n_1\bar{V}_1 + n_2\bar{V}_2)$ allows Equation 242 to be simplified to the desired relation,

$$\left(\frac{\partial^2 A}{\partial C^2}\right)_{T,V} = -\frac{\delta V}{C\bar{V}_1}\left(\frac{\partial \mu_1}{\partial C}\right)_{T,V} \qquad (243)$$

2. Virial Expansion for the Chemical Potential

In the limit of infinite dilution, solutions tend to become ideal and the following expression for the concentration dependence of the chemical potential can be used.

$$\mu_1 - \mu_1^0 = RT \ln X_1 \tag{244}$$

Here μ_1 is the chemical potential of species 1, $\mu_1°$ is the standard chemical potential of species 1, X_1 is the mole fraction of species 1, and R and T have their usual meaning. For a binary system it is clear that $X_1 = 1 - X_2$ so that

$$\mu_1 - \mu_1^0 = RT \ln (1 - X_2) \tag{245}$$

where X_2 is the mole fraction of the solute and can be approximated by $C\,\bar{V}_1/M$. Since the expansion of $\ln(1 - X_2)$ is given by

$$\ln(1 - X_2) = -X_2 - \frac{1}{2} X_2^2 - \ldots \tag{246}$$

Equation 245 can be rewritten as

$$\mu_1 - \mu_1^0 = -RTV_1^0\, C \left[\frac{1}{M} + \frac{V_1^0 C}{2M^2}\right] \tag{247}$$

It should be noted that at the limit of C = O, the partial molar volume $\bar{V}_1$ becomes the molar volume $V°_1$. Equation 247 is of the form of the well-known virial expansion which is usually written as

$$\mu_1 - \mu_1^0 = -RT\, V_1^0\, C \left[\frac{1}{M} + B_2 C + B_3 C^2 + \ldots\right] \tag{248}$$

where B_n is the nth virial coefficient. Equation 248 is differentiated with respect to solute concentration to obtain

$$\left(\frac{\partial \mu_1}{\partial C}\right)_{T,V} = -RT\, \bar{V}_1 \left[\frac{1}{M} + 2B_2 C + 3B_3 C^2 + \ldots\right] \tag{249}$$

Using the fact that $R = N_A k_B$ and rearranging, the standard form of the concentration dependence of the chemical potential is obtained.

$$-\frac{1}{k_B T \bar{V}_1}\left(\frac{\partial \mu_1}{\partial C}\right)_{T,V} = N_A\left[\frac{1}{M} + 2B_2C + 3B_3C^2 + \ldots\right] \tag{250}$$

D. Number, Weight, and z-Averages

In the study of macromolecules, polydisperse samples are often encountered which present distributions of values of properties such as molecular weight, radius of gyration, and the degree of polymerization.[110] Different analytical methods report different types of averages for these properties. For example, in osmotic pressure measurements, the equilibrium across the membrane is influenced by the concentration of particles. The molecular weight determined by this method is the number average molecular weight, which is defined by

$$M_n = \frac{\sum_i N_i M_i}{\sum_i N_i} \tag{251}$$

where M_i is the molecular weight of the ith species and N_i is the number of particles of the ith type per unit volume.

On the other hand, analytic methods such as light scattering and ultracentrifugation give averages that depend on the weight of the macromolecule. The weight average molecular weight is defined by

$$M_w = \frac{\sum_i (N_i M_i) M_i}{\sum N_i M_i} \tag{252}$$

That the weight average molecular weight is obtained from light scattering data when the scattering intensities are plotted as $(KC/R_\theta)_{c=0}$ vs. $\sin^2(\theta_s/2)$ can be seen from the following. First, assume that each species present has the same optical constant K and that the total concentration is given by

$$C = \sum_i C_i$$

At very low concentrations we can rewrite Equation 32 as

$$R_\theta = K \sum_i C_i M_i \tag{253}$$

Then using the identity $C_i = M_i N_i / N_A$, we find that

$$\left(\frac{KC}{R_\theta}\right)_{C=O} = \frac{\sum_i C_i}{\sum_j C_i M_i} = M_w^{-1} \tag{254}$$

The weight average molecular weight tends to emphasize the higher molecular weight species.

The molecular weights obtained from different averages may be quite different, depending on the distribution of the solute mass. Of course for a monodisperse solution the averages will be identical. This fact leads to an important property of these averages in that the degree of polydispersity of the system can be accessed. When a distribution of molecular weights is present, the various averages will assume a definite numerical order, i.e.,

$$M_n < M_w < M_z < M_{z+1} \cdots$$

where M_z, the z-average molecular weight, is defined as

$$M_z = \frac{\sum_i (N_i M_i^2) M_i}{\sum_i N_i M_i^2} \tag{255}$$

For a polydisperse sample the initial slope of a plot of $(K C/R_\theta)$ vs. $\sin^2(\theta_s/2)$ turns out to give the z-average radius of gyration.

$$S(\underset{\sim}{K})_{exp} = \frac{R_\theta}{R_\theta (\text{at } \theta_s = O)} = \frac{\sum_i S(\underset{\sim}{K})_i C_i M_i}{\sum_i C_i M_i} \tag{256}$$

where $S(\underset{\sim}{K})_i$ is the structure factor for the ith species. Using Equation 57 the inverse of the experimental structure factor can be expressed as

$$S(\underset{\sim}{K})^{-1} = \sum_i C_i M_i \left[1 - \frac{16\pi^2 n^2 (R_G^2)_i \sin^2(\theta_s/2)}{3\lambda_o^2}\right] \Big/ \sum_i C_i M_i \tag{257}$$

Then, since

$$<R_G^2>_z = \frac{\sum_i C_i M_i (R_G^2)_i}{\sum_i C_i M_i} = \frac{\sum_i N_i M_i^2 (R_G^2)_i}{\sum_i N_i M_i^2} \tag{258}$$

and the second term in the brackets is small

$$S(\underset{\sim}{K})^{-1} = 1 + \frac{16\pi^2 n^2}{3\lambda_0^2} <R_G^2>_z \sin^2(\theta_s/2) \quad (259)$$

Equation 259 can be combined with Equation 58 to give the desired result

$$\left(\frac{KC}{R_\theta}\right)_{C=O} = \frac{1}{M_w}\left[1 + \frac{16\pi^2 n^2 <R_G^2>_z}{3\lambda_0^2} \sin^2\left(\frac{\theta_s}{2}\right)\right] \quad (260)$$

E. Correlation Functions and Spectra of Scattered Light

Suppose the x(t) is a random variable, i.e., a function of t with an allowed set of values, each of which occurs with a definite probability. The power associated with x(t) is $|x(t)|^2$, and the power spectrum $J(\omega)$ is defined so that $J(\omega)d\omega$ gives the fraction of the average power which lies in the frequency range from ω to $\omega + d\omega$. The correlation function $C(\tau)$ for x(t) is written as[111,112]

$$C(\tau) = <x^*(t)x(t+\tau)> = \lim_{T\to\infty} \frac{1}{T} \int_{-T/2}^{+T/2} x^*(t)x(t+\tau)dt \quad (261)$$

For example x(t) might be the noise on an electrical signal, e.g., current fluctuations in a resistor. In this case $J(\omega)$ would be proportional to the signal from a narrow band amplifier tuned to detect the component of x(t) at ω. The purpose of this appendix is to derive the Wiener-Khintchine theorem which relates $J(\omega)$ to $C(\tau)$.

The record of x(t) in the time interval $-T/2 \leqslant t \leqslant T/2$ can be Fourier analyzed to obtain the frequency components of x(t). For this purpose we define $x_T(t)$ so that

$$x_T(t) = x(t) \quad \text{for} \quad |t| \lessgtr T/2$$

$$x_T(t) = 0 \text{ otherwise}$$

The variable $x_T(t)$ and its Fourier transform $\hat{x}_T(\omega)$ are related by the transform pair

$$x_T(t) = \int_{-\infty}^{+\infty} \hat{x}_T(\omega)e^{-i\omega t}\, d\omega \quad (262)$$

$$\hat{x}_T(\omega) = \frac{1}{2\pi} \int_{-T/2}^{+T/2} x_T(t)e^{+i\omega t}\, dt \quad (263)$$

First, we obtain an expression for $J(\omega)$ in terms of $\hat{x}_T(\omega)$ which is consistent with the identity

$$< | x_T(t) |^2 > = \int_{-\infty}^{+\infty} J(\omega) d\omega \tag{264}$$

The average power can be written as

$$< | x_T(t) |^2 > = \lim_{T\to\infty} \frac{1}{T} \int_{-T/2}^{+T/2} x_T^*(t) x_T(t) dt \tag{265}$$

and this provides a starting point. By inserting the rhs of Equation 262 for $x_T(t)$ into the integral of Equation 265 we obtain

$$\int_{-T/2}^{+T/2} x_T^*(t) x_T(t) dt = \int_{-T/2}^{+T/2} x_T^*(t) \left[\int_{-\infty}^{+\infty} \hat{x}_T(\omega) e^{-i\omega t} d\omega \right] dt \tag{266}$$

Then by exchanging the order of integration and using the complex conjugate of Equation 263, we find that

$$\int_{-\infty}^{+\infty} \hat{x}_T(\omega) \left[\int_{-T/2}^{+T/2} x_T^*(t) e^{-i\omega t} dt \right] d\omega = \int_{-\infty}^{+\infty} \hat{x}_T(\omega) [2\pi x_T^*(\omega)] d\omega \tag{267}$$

A comparison of Equations 267 and 265 then shows that

$$< | x_T(t) |^2 > = \int_{-\infty}^{+\infty} \lim_{T\to\infty} \left(\frac{2\pi |\hat{x}_T(\omega)|^2}{T} \right) d\omega \tag{268}$$

and Equation 264 requires that

$$J(\omega) = \lim_{T\to\infty} \frac{(2\pi) |\hat{x}_T(\omega)|^2}{T} \tag{269}$$

Using the form of $J(\omega)$ given in Equation 269, it is easy to show that the power spectrum is equal to the Fourier transform of the correlation function. With $\hat{x}_T(\omega)$ from Equation 263 we obtain

$$J(\omega) = \lim_{T\to\infty} \frac{2\pi}{T} \left[\frac{1}{2\pi}\int_{-T/2}^{+T/2} x_T^*(t)e^{-i\omega t}dt\right]\left[\frac{1}{2\pi}\int_{-T/2}^{+T/2} x_T(t')e^{+i\omega t'}dt'\right]$$

$$= \lim_{T\to\infty} \frac{1}{2\pi T} \int_{-T/2}^{+T/2}\int_{-T/2}^{+T/2} x_T^*(t)x_T(t')e^{-i\omega(t-t')}dtdt' \qquad (270)$$

Changing the variables of integration from t, t′ to t, τ where $\tau = t' - t$ gives

$$J(\omega) = \frac{1}{2\pi}\int_{-T/2}^{+T/2} \left[\lim_{T\to\infty} \frac{1}{T}\int_{-T/2}^{+T/2} x_T^*(t)x_T(t+\tau)dt \quad e^{i\omega\tau}\, d\tau\right] \qquad (271)$$

Since in cases of interest $C(\tau)$ vanishes for $\tau > T/2$, this is equivalent to

$$J(\omega) = \frac{1}{2\pi}\int_{-\infty}^{+\infty} C(\tau)e^{i\omega\tau}d\tau \qquad (272)$$

and the reverse transform is

$$C(\tau) = \int_{-\infty}^{+\infty} J(\omega)e^{-i\omega\tau}d\omega \qquad (273)$$

Equation 272 can be written in a form involving only integration over positive values of τ by considering the properties of the correlation function $C(\tau)$. Since x(t) represents a stationary random variable, any time origin can be used in the calculation of $C(\tau)$. Therefore

$$C(\tau) = \langle x^*(\tau) \;\; x(t+\tau)\rangle = \langle x^*(t-\tau) \;\; x(\tau)\rangle = C^*(-\tau) \qquad (274)$$

The identity $C(\tau) = C^*(-\tau)$ could also have been derived from the requirement that $J(\omega)$ be a real function, i.e., that $J(\omega) = J^*(\omega)$. In order to use this condition, we write Equation 272 as

$$J(\omega) = \frac{1}{2\pi}\left[\int_{-\infty}^{0} C(\tau)\, e^{i\omega\tau}d\tau + \int_{0}^{\infty} C(\tau)\, e^{i\omega\tau}\, d\tau\right] \qquad (275)$$

The first integral on the rhs of Equation 275 can be combined with 274 to obtain:

$$\int_{-\infty}^{0} C^* (-\tau) e^{i\omega\tau} d\tau = \int_{0}^{\infty} C^*(\tau)\, e^{-i\omega\tau} d\tau \tag{276}$$

Therefore Equation 275 can be written as

$$J(\omega) = \frac{1}{2\pi} \int_{0}^{\infty} [C^*(\tau) e^{-i\omega\tau} + C(\tau) e^{+i\omega\tau}]\, d\tau \tag{277}$$

and

$$J(\omega) = \frac{1}{\pi} \operatorname{Re} \int_{0}^{\infty} C(\tau)\, e^{i\omega\tau} d\tau \tag{278}$$

The last step follows from the identity $\operatorname{Re} z = (z + z^*)/2$ where z is an arbitrary random variable.

F. The Heterodyne Correlation Function

If I_{det} is the intensity of light at the surface of a PMT, the photocurrent or the photon count rate can be used to calculate an approximation to $G^{(2)}(\tau) = \langle I_{det}(t)\, I_{det}(t + \tau)\rangle$. In "homodyne" or "self-beat" experiments $I_{det}(t) = I_s$ and $G^{(2)}(\tau) = \langle I_s(t)\, I_s(t + \tau)\rangle$. As pointed out in Section III.A.3, this function does not in general permit the electric field correlation function $G^{(1)}(\tau)$ to be determined. Even with Gaussian light, only $|G^{(1)}(\tau)|$ can be derived from $G^{(2)}(\tau)$, as discussed in Section III.A.3, and important oscillatory factors may be lost. It is possible, however, to arrange the experiment so that $\operatorname{Re} G^{(1)}(\tau)$ is obtained directly. A factor such as $e^{i\omega_1 t}$, which would be lost if the absolute magnitude were taken, will then appear as $\cos(\omega_1 t)$. For this to be possible, a coherent reference beam of light must be mixed with the scattered light on the surface of the PMT. An analysis of this experiment follows.[37]

Suppose that a reference beam is obtained either by picking off a fraction of the incident beam before it reaches the scattering volume or by inserting an object into the scattering volume which will scatter incident light without a frequency shift. The latter method might be realized by simply collecting some light scattered from the wall of the sample cell in addition to light scattered from the sample. By analogy to radio frequency techniques, this is called a heterodyne experiment, and the reference beam is referred to as the local oscillator. The electric field of the light at the PMT is then given by

$$E_{det}(t) = E_S(t) + E_L(t) \tag{279}$$

where

$$E_S(t) = f(t)\, E_0 e^{-i\omega_0 t}$$

$$E_L(t) = E_{LO}\, e^{-i[\omega_0 t - \phi_L]}$$

Here f(t) specifies the modulation resulting from the scattering process, and ϕ_L takes into account a phase shift for E_L if the reference beam has a different path length to the detector than does the scattered beam. The autocorrelation function for I_{det} in this case is

$$G^{(2)}(\tau) = <[E_L^*(t) + E_S^*(t)]\; [E_L(t) + E_S(t)]\; [E_L^*(t+\tau) + E_S^*(t+\tau)]\, [E_L(t+\tau) + E_S(t+\tau)]>$$

This function can be multiplied out and simplified to obtain

$$G^{(2)}(\tau) = I_L^2 + 2\, I_L <I_S> + <I_S>^2 + I_L <E_S^*(t)\, E_S(t+\tau)>\, e^{+i\omega_0 \tau} + I_L <E_S(t)\, E_S^*(t+\tau)>\, e^{-i\omega_0 \tau} \tag{280}$$

where we have used the following results

$$<E_L(t)> = <E_S(t)> = <E_S(t)\, E_S(t+\tau)> = <E_L(t)\, F_L(t+\tau)> = 0$$

and similar equations for the complex conjugates. All of these averages vanish because

$$<e^{i\omega_o t}> = <\cos \omega_0 t> + i <\sin \omega_0 t> = 0$$

As in Section III, we suppress the factor of $(\varepsilon_o c/2)$ so that $I_L = |E_L|^2$ and $<I_s> = <|E_s(t)|^2>$. In the experiment, as usually performed, the intensity of the reference beam is adjusted so that $I_{LO} >> <I_s>$. Therefore, Equation 280 can be written as

$$G^{(2)}(\tau) = I_L^2 + 2\, I_L\, \mathrm{Re}[G_s^{(1)}(\tau) e^{+i\omega_0 \tau}] \tag{281}$$

According to Equation 273, the power spectrum associated with I_{det} in the heterodyne experiment is

$$I^{(2)}(\omega) = I_L^2\, \delta(\omega) + 2\, I_L \frac{\mathrm{Re}}{\pi} \int_0^{+\infty} e^{i\omega\tau}\, \mathrm{Re}[G_s^{(1)}(\tau) e^{i\omega_0 \tau}]\, d\tau \tag{282}$$

It is also possible to use a local oscillator having a frequency ω_L different from ω_o. In this case ω_o appearing in Equations 280, 281, and 282 would be replaced with ω_L. A discussion of this experiment and other details can be found in References 37 and 113.

G. Cumulant Analysis

The homodyne light scattering experiment yields $\beta|g^{(1)}(\tau)|^2$, and for monodisperse scatterers $|g^{(1)}(\tau)| = \exp(-\Gamma\tau)$ where $\Gamma = D_T K^2$. However, samples of macromolecules are often polydisperse because of aggregation or contamination, and a distribution of Γ values must be considered. If $G(\Gamma)$ is the normalized distribution function for Γ, then

$$|g^{(1)}(\tau)| = \int_0^\infty G(\Gamma)\, e^{-\Gamma\tau}\, d\Gamma \tag{283}$$

and

$$\int_0^\infty G(\Gamma)\, d\Gamma = 1$$

Discrete particle sizes can be treated by writing $G(\Gamma)$ in the form

$$G(\Gamma) = \frac{\sum_i N_i m_i^2\, \delta(\Gamma - \Gamma_i)}{\sum_i N_i m_i^2} = \frac{\sum_i C_i M_i\, \delta(\Gamma - \Gamma_i)}{\sum_i C_i M_i} \tag{284}$$

The notation is consistent with that used in Section II.1. For the ith species N_i is the number of scatterers per unit volume, m_i is the mass of a particle, C_i is the weight concentration (weight/unit volume), and M_i is the molecular weight. The form of $G(\Gamma)$ in Equation 284 shows that the average specified in Equation 283 is a z-average.

One method for obtaining the cumulant expansion of $\ln|g^{(1)}(\tau)|$ is to expand $\exp(-\Gamma\tau)$ about $\exp(-\langle\Gamma\rangle\tau)$. Thus[40]

$$\begin{aligned} |g^{(1)}(\tau)| &= e^{-\langle\Gamma\rangle\tau} \int_0^\infty G(\Gamma)\, e^{-(\Gamma-\langle\Gamma\rangle)\tau}\, d\Gamma \\ &= e^{-\langle\Gamma\rangle\tau} \int_0^\infty G(\Gamma)\left[1 - (\Gamma - \langle\Gamma\rangle)\tau + (\Gamma - \langle\Gamma\rangle)^2 \frac{\tau^2}{2!} + \ldots\right] d\Gamma \end{aligned} \tag{285}$$

where

$$<\Gamma> = \int_0^\infty \Gamma G(\Gamma)d\Gamma$$

The moments of the distribution are defined by

$$\mu_n = \int_0^\infty (\Gamma - <\Gamma>)^n G(\Gamma)d\Gamma \tag{286}$$

and Equation 285 can be written as

$$|g^{(1)}(\tau)| = e^{-<\Gamma>\tau}\left[1 + \frac{\mu_2}{2!}\tau^2 - \frac{\mu_3}{3!}\tau^3 + \frac{\mu_4}{4!}\tau^4 + \ldots\right] \tag{287}$$

The desired result is obtained using the Taylor's series

$$\ell n(1+x) = x - \frac{x^2}{2} + \frac{x^3}{3} - \frac{x^4}{4} + \ldots, \quad (-1<x<+1) \tag{288}$$

and collecting terms so that

$$\ell n|g^{(1)}(\tau)| = -<\Gamma>\tau + \frac{\mu_2}{2!}\tau^2 - \frac{\mu_3}{3!}\tau^3 + \frac{(\mu_4 - 3\mu_2^2)}{4!}\tau^4 + \ldots \tag{289}$$

The coefficients in this series, which are known as cumulants, describe some of the properties of G(Γ). The standard deviation is $\sqrt{\mu_2}$, the skewness is $\mu_3/\mu_2^{3/2}$, and the kurtosis is $[(\mu_4/\mu_2^2) - 3]/2$. Another quantity, which is often used to specify the degree of polydispersity, is the normalized variance defined as $\mu_2/<\Gamma>^2$. Since $\Gamma = D_T K^2$, the variance can also be expressed as

$$\frac{<(\Gamma - <\Gamma>)^2>}{<\Gamma>^2} = \frac{<D_T^2> - <D_T>^2}{<D_T>^2} \tag{290}$$

H. The Diffusion Coefficient

An expression for D_T can easily be obtained by considering the particle flux $\underset{\sim}{J}(\underset{\sim}{r},t)$.[42] From the definition of the flux as mass through unit area in unit time we can write

$$\underset{\sim}{J}(\underset{\sim}{r},t) = \underset{\sim}{v}(\underset{\sim}{r},t)\, C(\underset{\sim}{r},t) \tag{291}$$

where $\underset{\sim}{v}$ is the average velocity of the solute molecules at the position $\underset{\sim}{r}$ at time t, and

C is again the mass concentration of the solute. We assume that $\tilde{v}$ is a steady-state velocity which is attained by a particle under the influence of an applied force $\tilde{F}$ and a frictional force $-f_T\tilde{v}$ where f_T is the translational friction coefficient. For a sphere in a viscous medium Stokes' equation gives

$$f_T = 6\pi\eta a_h \tag{292}$$

where η is the coefficient of viscosity and a_h is the hydrodynamic radius of the sphere. Perrin and others have extended Stokes' treatment to cover ellipsoids of revolution. If the semi-axes are denoted by a, b, and b, Equation 292 can be rewritten as[114]

$$f_T = \frac{6\pi\eta a}{G(b/a)} \tag{293}$$

where

$$G(b/a) = \frac{1}{\sqrt{1-(b/a)^2}} \ln\left[\frac{1+\sqrt{1-(b/a)^2}}{(b/a)}\right] \quad ; \ a > b \text{ (prolate)}$$

and

$$G(b/a) = \frac{\tan^{-1}\sqrt{(b/a)^2-1}}{\sqrt{(b/a)^2-1}} \quad ; \ a < b \text{ (oblate)}$$

According to Newton's second law for a particle of mass m

$$m\frac{d^2\tilde{r}}{dt^2} = \text{net force} = \tilde{F} - f_T\tilde{v} \tag{294}$$

and under steady-state conditions, where the acceleration vanishes, we find that

$$\tilde{v} = \tilde{F}/f_T \tag{295}$$

In translational diffusion the effective force per particle results from the gradient in the chemical potential. Therefore, the driving force for diffusion can be written as

$$\tilde{F} = -\frac{1}{N_A}\tilde{\nabla}\mu_2 \tag{296}$$

where μ_2 is the chemical potential of the solute. Equations 291, 295, and 296 can be combined to obtain

$$\underset{\sim}{J}(\underset{\sim}{r},t) = -\frac{C(\underset{\sim}{r},t)}{f_T N_A}\underset{\sim}{\nabla}\mu_2 \tag{297}$$

In order to obtain an expression for D_T by comparing Equations 297 and 96 we must express $\underset{\sim}{\nabla}\mu_2$ in terms of the concentration. The chemical potential μ_2 is related to the mass concentration by the equation

$$\mu_2 = \mu_2^\circ + RT\,\ell n\,[\gamma C(\underset{\sim}{r}, t)/M] \tag{298}$$

where μ°_2 is the standard chemical potential, which is a constant, and γ is the activity coefficient. Thus

$$\underset{\sim}{\nabla}\mu_2 = \frac{RT}{C(r,t)}\underset{\sim}{\nabla}C(\underset{\sim}{r},t) \tag{299}$$

and

$$\underset{\sim}{J}(\underset{\sim}{r},t) = -\frac{RT}{N_A f_T}\underset{\sim}{\nabla}C(\underset{\sim}{r},t) \tag{300}$$

Equations 300 and 96 can now be combined to obtain the Stokes-Einstein equation

$$D_T = \frac{k_B T}{f_T} = \frac{k_B T}{6\pi\eta a_h} \tag{301}$$

We choose to retain the form of the diffusion equation as the concentration increases and simply to let D_T become a function of C. The concentration dependence of D_T enters the theory both through the chemical potential μ_2 and the friction coefficient f_T. From Appendix C we have the virial expansion for the chemical potential μ_1 of the solvent

$$\mu_1 = \mu_1^\circ - RT\,V_1^\circ\,C\left[\frac{1}{M} + B_2\,C + B_3\,C^2 + \ldots\right]$$

where V°_1 is the molar volume of the solvent and the B_i are virial coefficients. Considering only the x-direction, we have

$$\frac{\partial\mu_1}{\partial x} = -RT\,V_1^\circ\left(\frac{1}{M} + 2B_2\,C + 3B_3\,C^2 + \ldots\right)\frac{\partial C}{\partial x} \tag{302}$$

The desired expression for the gradient of μ_2 is obtained by using the Gibbs-Duhem relation $n_1 d\mu_1 = -n_2 d\mu_2$ where n_1 and n_2 are the number of moles of solvent and solute, respectively. Equation 302 then becomes

$$\frac{\partial \mu_2}{\partial x} = \frac{RT}{C} (1 + 2B_2 MC + 3B_3 MC^2 + \ldots) \frac{\partial C}{\partial x} \quad (303)$$

where we have used the fact that $(n_1 V^\circ{}_1/n_2 M) = C^{-1}$. The friction coefficient can also be expanded in terms of C as follows

$$f_T = f_o (1 + B'C + \ldots) \quad (304)$$

When Equations 303 and 304 are combined with Equation 297, the resulting expression for D_T is

$$\begin{aligned} D_T(C) &= \frac{k_B T}{f_o} \frac{(1 + 2\,B_2\,MC + 3\,B_3\,MC^2 + \ldots)}{(1 + B'C + \ldots)} \\ &= \frac{k_B T}{f_o} [1 + (2\,M\,B_2 - B')\,C + \ldots] \end{aligned} \quad (305)$$

I. The Rotational Diffusion Equation

The rotational random walk of a unit vector is depicted in Figure 19. By considering the probability of rotations into and out of an element of solid angle $d\Omega$, Debye was able to derive an equation for rotational diffusion similar to the translational diffusion equation. The conditional probability that the orientation is along r at time τ given that it was along $\hat{r}_o$ at $t = 0$ is denoted by $P(\hat{r}_o|\hat{r},\tau)$, which is also known as the transition probability from $\hat{r}_o$ to $\hat{r}$. This function obeys the rotational diffusion equation[77]

$$\frac{\partial P(\hat{r}_o|\hat{r},\tau)}{\partial \tau} = D_R \nabla^2 P(\hat{r}_o|\hat{r},\tau) \quad (306)$$

where in spherical polar coordinates

$$\nabla^2 = \frac{1}{\sin\theta} \frac{\partial}{\partial\theta} \left(\sin\theta \frac{\partial}{\partial\theta} \right) + \frac{1}{\sin^2\theta} \frac{\partial^2}{\partial\phi^2} \quad (307)$$

In the following we take the polar coordinates of $\hat{r}_o$ and $\hat{r}$ to be (θ_o,ϕ_o) and (θ,ϕ), respectively.

Students of atomic physics will recognize that the angular momentum operator $\hat{L}^2$ is equal to the negative of ∇^2. In addition it is well known that the spherical harmonics $Y_{\ell m}(\theta,\phi)$ are eigenfunctions of $\hat{L}^2$ and satisfy the equation[115]

$$\hat{L}^2 Y_{\ell m}(\theta,\phi) = \ell(\ell+1)\, Y_{\ell m}(\theta,\phi) \tag{308}$$

where $\ell = 0,1,2,3 \ldots, m = \ell, \ell-1, \ell-2, \ldots, -\ell+1, -\ell$. Therefore,

$$\nabla^2 Y_{\ell m}(\theta,\phi) = -\ell(\ell+1)\, Y_{\ell m}(\theta,\phi) \tag{309}$$

Using this result, it is easy to verify that the following is a solution of Equation 306.

$$P(\hat{r}_0|\hat{r},\tau) = Y^*_{\ell m}(\theta,\phi)\, e^{-\ell(\ell+1) D_R \tau} \tag{310}$$

A linear combination of solutions is also a solution, and any solution can be written as

$$P(r_0|r,\tau) = \sum_{\ell=0}^{\infty} \sum_{m=-\ell}^{\ell} c_{\ell m}\, Y^*_{\ell m}(\theta,\phi)\, e^{-\ell(\ell+1) D_R \tau} \tag{311}$$

where the $c_{\ell m}$ are coefficients which must be determined. The functions $Y_{\ell m}(\theta,\phi)$ are a complete set of normalized and orthogonal functions. Thus

$$\int_{\phi=0}^{2\pi} \int_{\theta=0}^{\pi} Y^*_{\ell m}(\theta,\phi)\, Y_{\ell' m'}(\theta,\phi) \sin\theta\, d\theta\, d\phi = \delta_{\ell\ell'}\, \delta_{mm'} \tag{312}$$

where the δ_{ij} are Kronecker delta functions which are unity for $i = j$ and are zero otherwise. Therefore, $c_{\ell m}$ can be determined by setting $\tau = 0$ and using the usual trick of multiplying through Equation 311 by $Y_{\ell m}(\theta,\phi)$ and integrating over the coordinates. We recognize that

$$P(\hat{r}_0|\hat{r},0) = \delta(\hat{r}_0 - \hat{r}) = \delta(\theta - \theta_0)\, \delta(\phi - \phi_0) \tag{313}$$

where $\delta(x - x_0)$ is the Dirac delta function which is normalized but equals zero if $x \neq x_0$. The integration then gives

$$\int Y_{\ell m}(\theta,\phi)\, P(\hat{r}_0|\hat{r},0)\, d\Omega = c_{\ell m} = Y_{\ell m}(\theta_0,\phi_0) \tag{314}$$

Now returning to Equation 311 we can write

$$P(r_0|r,\tau) = \sum_{\ell=0}^{\infty} \sum_{m=-\ell}^{\ell} Y_{\ell m}(\theta_0,\phi_0)\, Y^*_{\ell m}(\theta,\phi)\, e^{-\ell(\ell+1) D_R \tau} \tag{315}$$

It is this probability function which is required in the calculation of correlation functions of the components of the polarizability tensor.

REFERENCES

1. Tanford, C., *Physical Chemistry of Macromolecules,* John Wiley & Sons, New York, 1961.
2. Timasheff, S. N. and Townsend, R., Light scattering, *Physical Principles and Techniques of Protein Chemistry Part B,* Leach, S. J., Ed., Academic Press, New York, 1970, 147.
3. Huglin, M. B., *Light Scattering from Polymer Solutions,* Academic Press, New York, 1972.
4. Fabelinskii, I. L., *Molecular Scattering of Light,* Plenum Press, New York, 1968.
5. Kerker, M., *The Scattering of Light,* Academic Press, New York, 1969.
6. Long, D. A., *Raman Spectroscopy,* McGraw-Hill, New York, 1977.
7. Schurr, J. M., Dynamic light scattering of biopolymers and biocolloids, *CRC Crit. Rev. Biochem.,* 4, 371, 1977.
8. Pusey, P. N. and Vaughan, J. M., Light scattering and intensity fluctuation spectroscopy, *Dielectric and Related Molecular Processes,* Vol. 2, Davies, M., Ed., The Chemical Society, London, 1975, 48.
9. Carlson, F. D., The application of intensity fluctuation spectroscopy to molecular biology, *Ann. Rev. Biophys. and Bioeng.,* 4, 243, 1975.
10. Ware, B. R., Applications of laser velocimetry in biology and medicine, *Chemical and Biochemical Applications of Lasers,* Vol. 2, Moore, C. B., Ed., Academic Press, New York, 1977, chap. 5.
11. Ware, B. R., Electrophoretic light scattering, *Adv. Colloid Interface Sci.,* 4, 1, 1974.
12. Berne, B. J. and Pecora, R., *Dynamic Light Scattering,* John Wiley & Sons, New York, 1976.
13. Chu, B., *Laser Light Scattering,* Academic Press, New York, 1974.
14. Cummins, H. Z. and Pike, E. R., *Photon Correlation and Light Beating Spectroscopy,* Plenum Press, New York, 1974.
15. Loudon, R., *The Quantum Theory of Light,* Clarendon Press, Oxford, 1973.
16. Rayleigh, Lord, On the light from the sky, its polarization and color, *Phil. Mag.,* XLI, 4th series, 107, 1871.
17. Kaye, W. and Havlik, A. J., Low angle laser light scattering — absolute calibration, *Appl. Opt.,* 12, 541, 1973.
18. Smoluchowski, M., Molekular-kinetische Theorie der Opaleszenz von Gasen im kritischen Zustande sowie einiger verwandter Erscheinungen, *Ann. Phys.,* 25, 205, 1908.
19. Einstein, A., Theory of the opalescence of homogeneous liquids and mixtures of liquids in the vicinity of the critical state, English translation, *Colloid Chemistry,* Vol. I, Alexander, J., Ed., Reinhold, New York, 1926, 323.
20. Neugebauer, T., Berechnung der Lichtzerstreuung von Fadenkettenlosungen, *Ann. Phys.,* 42, 509, 1943.
21. Kratochvil, P., Particle scattering functions, *Light Scattering from Polymer Solutions,* Huglin, M. B., Ed., Academic Press, New York, 1972, chap. 7.
22. Gunier, A., La diffraction des rayons X aux tres petits angles: application a l'etude de phenomenes ultramicroscopique, *Ann. Phys.,* 12, 161, 1939.
23. Tanford, C., *Physical Chemistry of Macromolecules,* John Wiley & Sons, New York, 1961, chap. 3.
24. Zimm, B. H., Apparatus and methods for measurement and interpretation of angular variation of light scattering; preliminary results on polystyrene solutions, *J. Chem. Phys.,* 16, 1099, 1948.
25. Morris, V. J., Coles, H. J., and Jennings, B. R., Infrared plots for macromolecular characterization, *Nature,* 249, 240, 1974.
26. Benoit, H., On the effect of branching and polydispersity on the angular distribution of light scattering by Gaussian coils, *J. Polym. Sci.,* 11, 507, 1953.
27. Benoit, H., Holtzer, A. M., and Doty, P., An experimental study of polydispersity by light scattering, *J. Phys. Chem.,* 58, 635, 1954.

28. **Holtzer, A. M., Benoit, H., and Doty, P.**, The molecular configuration and hydrodynamic behavior of cellulose trinitrate, *J. Phys. Chem.*, 58, 624, 1954.
29. **Holtzer, A.**, Interpretation of the angular distribution of the light scattered by a polydisperse system of rods, *J. Polym. Sci.*, 17, 432, 1955.
30. **Goldstein, M.**, Scattering factors for certain polydisperse systems, *J. Chem. Phys.*, 21, 1255, 1953.
31. **Rice, S. A.**, Particle scattering factors in polydisperse systems, *J. Polym. Sci.*, 16, 94, 1955.
32. **Kratochvil, P.**, Some remarks on light scattering by solutions of highly polydisperse polymers, *J. Polym. Sci.*, Part C, 23, 143, 1968.
33. **Carpenter, D. K.**, Light scattering study of the molecular weight distribution of polypropylene, *J. Polym. Sci.*, Part A-2, 4, 923, 1966.
34. **Cummins, H. Z., Knable, N., and Yeh, Y.**, Observation of diffusion broadening of Rayleigh scattered light, *Phys. Rev. Lett.*, 12, 150, 1964.
35. **Mandel, L.**, Fluctuations in light beams, *Progress in Optics*, Vol. 2, Wolf, E., Ed., Wiley-Interscience, New York, 1963, 181.
36. **Pecora, R.**, Doppler shifts in light scattering from pure liquids and polymer solutions, *J. Chem. Phys.*, 40, 1604, 1964.
37. **Jakeman, E.**, Photon correlation, *Photon Correlation and Light Beating Spectroscopy*, Cummins, H. Z. and Pike, E. R., Eds., Plenum Press, New York, 1974, 75.
38. **Jones, C. R.**, Photon Correlation Spectroscopy of Hemoglobin: Diffusion of Oxy-HbA and Oxy-HbS, Ph.D. thesis, University of North Carolina, Chapel Hill, 1977.
39. **Wilson, W. W.**, Light Beating Spectroscopy Applied to Biochemical Systems, Ph.D. thesis, University of North Carolina, Chapel Hill, 1973.
40. **Koppel, D. E.**, Analysis of macromolecular dispersity in intensity correlation spectroscopy: the method of cumulants, *J. Chem. Phys.*, 57, 4814, 1972.
41. **Bezot, P., Ostrowsky, N., and Hesse-Bezot, C.**, Light scattering data analysis for samples with large polydispersities, *Optics Commun.*, 25, 14, 1978.
42. **Einstein, A.**, *Investigations on the Theory of the Brownian Movement*, Dover Publications, New York, 1926.
43. **Cummins, H. Z.**, Applications of light beating spectroscopy in biology, *Photon Correlation and Light Beating Spectroscopy*, Cummins, H. Z. and Pike, E. R., Eds., Plenum Press, New York, 1974, 285.
44. **Kitchen, R. G., Preston, B. N., and Wells, J. D.**, Diffusion and sedimentation of serum albumin in concentrated solutions, *J. Polymer Sci. Symp.*, 55, 39, 1976.
45. **Keller, K. H., Canales, E. R., and Yum, S. I.**, Tracer and mutual diffusion coefficients of proteins, *J. Phys. Chem.*, 75, 379, 1971.
46. **Herbert, T. J. and Carlson, F. D.**, Spectroscopic study of the self-association of myosin, *Biopolymers*, 10, 2231, 1971.
47. **Palmer, G., Fritz, O. G., and Hallett, F. R.**, Quasi-elastic light scattering on human fibrinogen. I. Fibrinogen, *Biopolymers*, 18, 1647, 1979.
48. **Cohen, R. J., Jedziniak, J. A., and Benedek, G. B.**, Study of the aggregation and allosteric control of bovine glutamate dehydrogenase by means of quasi-elastic light scattering spectroscopy, *Proc. R. Soc. London, Ser. A*, 345, 73, 1975.
49. **Jones, C. R., Johnson, C. S., Jr., and Penniston, J. T.**, Photon correlation spectroscopy of hemoglobin: diffusion of oxy-HbA and oxy-HbS, *Biopolymers*, 17, 1581, 1978.
50. **Dubin, S. B., Lunacek, J. H., and Benedek, G. B.**, Observation of the spectrum of light scattered by solutions of biological macromolecules, *Proc. Natl. Acad. Sci. U.S.A.*, 57, 1164, 1967.
51. **Dubin, S. B., Clark, N. A., and Benedek, G. B.**, Measurement of the rotational diffusion coefficient of lysozyme by depolarized light scattering: configuration of lysozyme in solution, *J. Chem. Phys.*, 54, 5158, 1971.
52. **Rimai, L., Hockmott, J. T., Jr., Cole, T., and Carew, E. B.**, Quasi-elastic light scattering by diffusional fluctuations in RNase solutions, *Biophys. J.*, 10, 20, 1970.
53. **Bellamy, A. R., Gillies, S. C., and Harvey, J. D.**, Molecular weight of two oncornavirus geomes: derivation from particle molecular weights and RNA content, *J. Virol.*, 14, 1388, 1974.
54. **Ware, B. R., Raj, T., Flygare, W. H., Lesnaw, J. A., and Reichman, M. E.**, Molecular weights of vesicular stomatitis virus and its defective particles by laser light-beating spectroscopy, *J. Virol.*, 11, 141, 1973.
55. **Schaefer, D. W., Benedek, G. B., Schofield, P., and Bradford, E.**, Spectrum of light quasielastically scattered from tobacco mosaic virus, *J. Chem. Phys.*, 55, 3884, 1971.
56. **Dubos, P., Hallett, R., Kells, D. T. C., Sorensen, O., and Rowe, D.**, Biophysical studies of infectious pancreatic necrosis virus, *J. Virol.*, 22, 150, 1977.

57. **Camerini-Otero, R. D., Pusey, P. N., Koppel, D. E., Schaefer, D. W., and Franklin, R. M.,** Intensity fluctuation spectroscopy of laser light scattered by solutions of spherical viruses: R17, Qβ, BSV, PM2, and T7. II. Diffusion coefficients, molecular weights, solvation, and particle dimensions, *Biochemistry,* 13, 960, 1974.

58. **Koppel, D. E.,** Study of *Escherichia coli* ribosomes by intensity fluctuation spectroscopy of scattered laser light, *Biochemistry,* 13, 2712, 1974.

59. **Tanford, C.,** *Physical Chemistry of Macromolecules,* John Wiley & Sons, New York, 1961, 356.

60. **Tanford, C.,** *Physical Chemistry of Macromolecules,* John Wiley & Sons, New York, 1961, 379.

61. **Tanaka, T., Riva, C., and Ben-Sira, I.,** Blood velocity measurements in human retinal vessels, *Science,* 186, 830, 1974.

62. **Tanaka, T. and Benedek, G. B.,** Measurement of the velocity of blood flow (in vivo) using a fiber optic catheter and optical mixing spectroscopy, *Appl. Opt.,* 14, 189, 1975.

63. **Mustacich, R. V. and Ware, B. R.,** Observation of protoplasmic streaming by laser-light scattering, *Phys. Rev. Lett.,* 33, 617, 1974.

64. **Ware, B. R. and Flygare, W. H.,** The simultaneous measurement of the electrophoretic mobility and diffusion coefficient in Bovine Serum albumin solutions by light scattering, *Chem. Phys. Lett.,* 12, 81, 1971.

65. **Tanford, C.,** *Physical Chemistry of Macromolecules,* John Wiley & Sons, New York, 1961, 414.

66. **Haas, D. D. and Ware, B. R.,** Design and construction of a new electrophoretic light scattering chamber and applications to solutions of hemoglobin, *Anal. Biochem.,* 74, 175, 1976.

67. **Moran, R., Steiner, R., and Kaufmann, R.,** Laser Doppler spectroscopy as applied to electrophoresis of protein solutions, *Anal. Biochem.,* 70, 506, 1976.

68. **Josefowicz, J. and Hallett, F. R.,** Homodyne electrophoretic light scattering of polystyrene spheres by laser cross-beam intensity correlation, *Appl. Optics,* 14, 740, 1975.

69. **Chu, B.,** *Laser Light Scattering,* Academic Press, New York, 1974, 283.

70. **Tanford, C.,** *Physical Chemistry of Macromolecules,* John Wiley & Sons, New York, 1961, 414.

71. **Uzgiris, E. E. and Kaplan, J. H.,** Study of lymphocyte and erythrocyte electrophoretic mobility by laser Doppler spectroscopy, *Anal. Biochem.,* 60, 455, 1974.

72. **Ware, B. R. and Flygare, W. H.,** Light scattering in mixtures of BSA, BSA dimers, and fibrinogen under the influence of electric fields, *J. Colloid Interface Sci.,* 39, 670, 1972.

73. **Luner, S. J., Szklarek, D., Knox, R. J., Seaman, G. V. F., Josefowicz, J. Y., and Ware, B. R.,** Red cell charge is not a function of cell age, *Nature (London),* 269, 719, 1977.

74. **Smith, B. A., Ware, B. R., and Weiner, R. S.,** Electrophoretic distributions of human peripheral blood mononuclear white cells from normal subjects and from patients with acute lymphocyte leukmia, *Proc. Natl. Acad. Sci. U.S.A.,* 73, 2388, 1976.

75. **Josefowicz, J. and Hallett, F. R.,** Cell surface effects of pokeweed observed by electrophoretic light scattering, *FEBS Lett.,* 60, 62, 1975.

76. **Rimai, L., Salmeen, I., Hart, D., Liebes, L., Rich, M. A., and McCormick, J. J.,** Electrophoretic mobilities of RNA tumor viruses. Studies by Doppler-shifted light scattering spectroscopy, *Biochemistry,* 14, 4621, 1975.

77. **Debye, P.,** *Polar Molecules,* Dover Publications, New York, 1929, chap. 5.

78. **Berne, B. J. and Pecora, R.,** *Dynamic Light Scattering,* John Wiley & Sons, New York, 1976, 143.

79. **Perrin, F.,** Movement Brownien d'un ellipsoide (II). Rotation libre et depolarization des fluorescences translation et diffusion de molecules ellipsoidales, *J. Phys. Radium,* 7, 1, 1936.

80. **Pecora, R.,** Spectral distribution of light scattered by monodisperse rigid rods, *J. Chem. Phys.,* 48, 4126, 1968.

81. **Cummins, H. Z., Carlson, F. D., Herbert, T. J., and Woods, G.,** Translational and rotational diffusion constants of tobacco mosaic virus from Rayleigh linewidths, *Biophys. J.,* 9, 518, 1969.

82. **Nossal, R.,** Spectral analysis of laser light scattered from motile microorganisms, *Biophys. J.,* 11, 341, 1971.

83. **Berne, B. J. and Pecora, R.,** *Dynamic Light Scattering,* John Wiley & Sons, New York, 1976, chap. 5.

84. **Nossal, R., Chen, S.-H., and Lai, C.-C.,** Use of laser scattering for quantitative determinations of bacterial motility, *Opt. Commun.,* 4, 35, 1971.

85. **Schaefer, D. W., Banks, G., and Alpert, S. S.,** Intensity fluctuation spectroscopy of motile microorganisms, *Nature (London),* 248, 162, 1974.

86. **Schaefer, D. W.,** Dynamics of number fluctuations: motile microorganisms, *Science,* 180, 1293, 1973.

87. **Schaefer, D. W. and Berne, B. J.,** Light scattered from non-Gaussian concentration fluctuations, *Phys. Rev. Lett.,* 28, 475, 1972.

88. **Magde, D., Elson, E., and Webb, W. W.,** Thermodynamic fluctuations in a reacting system-measurement by fluorescence correlation spectroscopy, *Phys. Rev. Lett.,* 29, 705, 1972.

89. **Feher, G. and Weissman, M.**, Fluctuation spectroscopy: determination of chemical reaction kinetics from the frequency spectrum of fluctuations, *Proc. Natl. Acad. Sci. U.S.A.*, 70, 870, 1973.
90. **Birch, A. D., Brown, D. R., Dodson, M. G., and Thomas, J. R.**, The determination of gaseous turbulent concentration fluctuations using Raman photon correlation spectroscopy, *J. Phys. D.*, 8, L167, 1975.
91. **Berne, B. J., Deutch, J. M., Hynes, J. T., and Frisch, H. L.**, Light scattering from chemically reactive mixtures, *J. Chem. Phys.*, 49, 2864, 1968.
92. **Feller, W.**, *An Introduction by Probability Theory and Its Applications*, Vol. 1, 3rd ed., John Wiley & Sons, New York, 1968, chap. 17.
93. **Bauer, D. R., Hudson, B., and Pecora, R.**, Resonance enhanced depolarized Rayleigh scattering from diphenylpolyenes, *J. Chem. Phys.*, 63, 588, 1975.
94. **Bloomfield, V. A. and Benbasat, J. A.**, Inelastic light-scattering study of macromolecular reaction kinetics. I. The reactions $A \rightleftharpoons B$ and $2A \rightleftharpoons A_2$, *Macromolecules*, 4, 609, 1971.
95. **Jakeman, E., Pusey, P. N., and Vaughan, J. M.**, Intensity fluctuation light-scattering spectroscopy using a conventional light source, *Optics Commun.*, 17, 305, 1976.
96. **Cummins, H. Z. and Swinney, H. L.**, Light beating spectroscopy, *Progress in Optics*, Vol. 8, Wolf, E., Ed., North-Holland, Amsterdam, 1970, 133.
97. **Mandel, L.**, Correlation properties of light scattered from fluids, *Phys. Rev.*, 181, 75, 1969.
98. **Jones, C. R. and Johnson, C. S., Jr.**, Photon correlation spectroscopy using a jet stream dye laser, *J. 'hem. Phys.*, 65, 2020, 1976.
99. **Gulari, E. and Chu, B.**, Photon correlation in the nanosecond range and its application to the evaluation of RCA C31034 photomultiplier tubes, *Rev. Sci. Instrum.*, 48, 1560, 1977.
100. **Lastovka, J. B.**, Light Mixing Spectroscopy and the Spectrum of Light Scattered by Thermal Fluctuations in Liquids, Ph.D. thesis, Massachusetts Institute of Technology, Cambridge, 1967.
101. **Jolly, D. and Eisenberg, H.**, Photon correlation spectroscopy, total intensity light scattering with laser radiation, and hydrodynamic studies of a well-fractionated DNA sample, *Biopolymers*, 15, 61, 1976.
102. **Kaye, W. and McDaniel, J. B.**, Low-angle laser light scattering-Rayleigh factors and depolarization ratios, *Appl. Optics*, 13, 1934, 1974.
103. **Gordon, J. P., Leite, R. C. C., Moore, R. S., Porto, S. P. S., and Whinnery, J. R.**, Long-transient effects in lasers with inserted liquid samples, *J. Appl. Phys.*, 36, 3, 1965.
104. **Whinnery, J. R.**, Laser measurement of optical absorption in liquids, *Acc. Chem. Res.*, 7, 225, 1974.
105. **Whinnery, J. R., Miller, D. T., and Dabby, F.**, Thermal convection and spherical aberration distortion of laser beams in low loss liquids, *IEEE J. Quantum Electron*, QE-3, 382, 1967.
106. **Carrington, A. and McLachlan, A. D.**, *Introduction to Magnetic Resonance*, Harper & Row, New York, 1967, 260.
107. **Long, D. A.**, *Raman Spectroscopy*, McGraw-Hill, New York, 1977, 46.
108. **Stacey, K. A.**, *Light Scattering in Physical Chemistry*, Academic Press, New York, 1956, 21.
109. **Loudon, R.**, *The Quantum Theory of Light*, Clarendon Press, Oxford, 1973, chap. 2.
110. **Tanford, C.**, *Physical Chemistry of Macromolecules*, John Wiley & Sons, New York, 1961, 145.
111. **Flygare, W. H.**, *Molecular Structure and Dynamics*, Prentice-Hall, Englewood Cliffs, N.J., 1978, chap. 1.
112. **McQuarrie, D. A.**, *Statistical Mechanics*, Harper & Row, New York, 1976, chap. 22.
113. **Chu, B.**, *Laser Light Scattering*, Academic Press, New York, 1974, 101.
114. **Perrin, F.**, Mouvement Brownien d'un ellipsoide (I.) Dispersion dielectrique pour des molecules ellipsoidales, *J. Phys. Radium*, 5, 497, 1934.
115. **Johnson, C. S., Jr. and Pedersen, L. G.**, *Problems and Solutions in Quantum Chemistry and Physics*, Addison-Wesley, Reading, Mass., 1974, chap. 5.
116. **Hall, R. S. and Johnson, C. S., Jr.**, Experimental evidence that mutual and tracer diffusion coefficients for hemoglobin are not equal, *J. Chem. Phys.*, 72, 4251, 1980.
117. **Hall, R. S., Oh, Y. S., and Johnson, C. S., Jr.**, Photon correlation spectroscopy in strongly absorbing and concentrated samples and applications to unliganded hemoglobin, *J. Phys. Chem.*, 84, 756, 1980.

A CATALOG OF SELECTED

DOVER BOOKS

IN SCIENCE AND MATHEMATICS

Astronomy

BURNHAM'S CELESTIAL HANDBOOK, Robert Burnham, Jr. Thorough guide to the stars beyond our solar system. Exhaustive treatment. Alphabetical by constellation: Andromeda to Cetus in Vol. 1; Chamaeleon to Orion in Vol. 2; and Pavo to Vulpecula in Vol. 3. Hundreds of illustrations. Index in Vol. 3. 2,000pp. $6^{1}/_{8}$ x $9^{1}/_{4}$.

Vol. I: 0-486-23567-X
Vol. II: 0-486-23568-8
Vol. III: 0-486-23673-0

EXPLORING THE MOON THROUGH BINOCULARS AND SMALL TELESCOPES, Ernest H. Cherrington, Jr. Informative, profusely illustrated guide to locating and identifying craters, rills, seas, mountains, other lunar features. Newly revised and updated with special section of new photos. Over 100 photos and diagrams. 240pp. $8^{1}/_{4}$ x 11. 0-486-24491-1

THE EXTRATERRESTRIAL LIFE DEBATE, 1750–1900, Michael J. Crowe. First detailed, scholarly study in English of the many ideas that developed from 1750 to 1900 regarding the existence of intelligent extraterrestrial life. Examines ideas of Kant, Herschel, Voltaire, Percival Lowell, many other scientists and thinkers. 16 illustrations. 704pp. $5^{3}/_{8}$ x $8^{1}/_{2}$. 0-486-40675-X

THEORIES OF THE WORLD FROM ANTIQUITY TO THE COPERNICAN REVOLUTION, Michael J. Crowe. Newly revised edition of an accessible, enlightening book re-creates the change from an earth-centered to a sun-centered conception of the solar system. 242pp. $5^{3}/_{8}$ x $8^{1}/_{2}$. 0-486-41444-2

ARISTARCHUS OF SAMOS: The Ancient Copernicus, Sir Thomas Heath. Heath's history of astronomy ranges from Homer and Hesiod to Aristarchus and includes quotes from numerous thinkers, compilers, and scholasticists from Thales and Anaximander through Pythagoras, Plato, Aristotle, and Heraclides. 34 figures. 448pp. $5^{3}/_{8}$ x $8^{1}/_{2}$.
0-486-43886-4

A COMPLETE MANUAL OF AMATEUR ASTRONOMY: TOOLS AND TECHNIQUES FOR ASTRONOMICAL OBSERVATIONS, P. Clay Sherrod with Thomas L. Koed. Concise, highly readable book discusses: selecting, setting up and maintaining a telescope; amateur studies of the sun; lunar topography and occultations; observations of Mars, Jupiter, Saturn, the minor planets and the stars; an introduction to photoelectric photometry; more. 1981 ed. 124 figures. 25 halftones. 37 tables. 335pp. $6^{1}/_{2}$ x $9^{1}/_{4}$. 0-486-42820-8

AMATEUR ASTRONOMER'S HANDBOOK, J. B. Sidgwick. Timeless, comprehensive coverage of telescopes, mirrors, lenses, mountings, telescope drives, micrometers, spectroscopes, more. 189 illustrations. 576pp. $5^{5}/_{8}$ x $8^{1}/_{4}$. (Available in U.S. only.)
0-486-24034-7

STAR LORE: Myths, Legends, and Facts, William Tyler Olcott. Captivating retellings of the origins and histories of ancient star groups include Pegasus, Ursa Major, Pleiades, signs of the zodiac, and other constellations. "Classic."—Sky & Telescope. 58 illustrations. 544pp. $5^{3}/_{8}$ x $8^{1}/_{2}$. 0-486-43581-4

Chemistry

THE SCEPTICAL CHYMIST: THE CLASSIC 1661 TEXT, Robert Boyle. Boyle defines the term "element," asserting that all natural phenomena can be explained by the motion and organization of primary particles. 1911 ed. viii+232pp. $5^3/_8 \times 8^1/_2$. 0-486-42825-7

RADIOACTIVE SUBSTANCES, Marie Curie. Here is the celebrated scientist's doctoral thesis, the prelude to her receipt of the 1903 Nobel Prize. Curie discusses establishing atomic character of radioactivity found in compounds of uranium and thorium; extraction from pitchblende of polonium and radium; isolation of pure radium chloride; determination of atomic weight of radium; plus electric, photographic, luminous, heat, color effects of radioactivity. ii+94pp. $5^3/_8 \times 8^1/_2$. 0-486-42550-9

CHEMICAL MAGIC, Leonard A. Ford. Second Edition, Revised by E. Winston Grundmeier. Over 100 unusual stunts demonstrating cold fire, dust explosions, much more. Text explains scientific principles and stresses safety precautions. 128pp. $5^3/_8 \times 8^1/_2$. 0-486-67628-5

MOLECULAR THEORY OF CAPILLARITY, J. S. Rowlinson and B. Widom. History of surface phenomena offers critical and detailed examination and assessment of modern theories, focusing on statistical mechanics and application of results in mean-field approximation to model systems. 1989 edition. 352pp. $5^3/_8 \times 8^1/_2$. 0-486-42544-4

CHEMICAL AND CATALYTIC REACTION ENGINEERING, James J. Carberry. Designed to offer background for managing chemical reactions, this text examines behavior of chemical reactions and reactors; fluid-fluid and fluid-solid reaction systems; heterogeneous catalysis and catalytic kinetics; more. 1976 edition. 672pp. $6^1/_8 \times 9^1/_4$. 0-486-41736-0 $31.95

ELEMENTS OF CHEMISTRY, Antoine Lavoisier. Monumental classic by founder of modern chemistry in remarkable reprint of rare 1790 Kerr translation. A must for every student of chemistry or the history of science. 539pp. $5^3/_8 \times 8^1/_2$. 0-486-64624-6

MOLECULES AND RADIATION: An Introduction to Modern Molecular Spectroscopy. Second Edition, Jeffrey I. Steinfeld. This unified treatment introduces upper-level undergraduates and graduate students to the concepts and the methods of molecular spectroscopy and applications to quantum electronics, lasers, and related optical phenomena. 1985 edition. 512pp. $5^3/_8 \times 8^1/_2$. 0-486-44152-0

A SHORT HISTORY OF CHEMISTRY, J. R. Partington. Classic exposition explores origins of chemistry, alchemy, early medical chemistry, nature of atmosphere, theory of valency, laws and structure of atomic theory, much more. 428pp. $5^3/_8 \times 8^1/_2$. (Available in U.S. only.) 0-486-65977-1

GENERAL CHEMISTRY, Linus Pauling. Revised 3rd edition of classic first-year text by Nobel laureate. Atomic and molecular structure, quantum mechanics, statistical mechanics, thermodynamics correlated with descriptive chemistry. Problems. 992pp. $5^3/_8 \times 8^1/_2$. 0-486-65622-5

ELECTRON CORRELATION IN MOLECULES, S. Wilson. This text addresses one of theoretical chemistry's central problems. Topics include molecular electronic structure, independent electron models, electron correlation, the linked diagram theorem, and related topics. 1984 edition. 304pp. $5^3/_8 \times 8^1/_2$. 0-486-45879-2

Engineering

DE RE METALLICA, Georgius Agricola. The famous Hoover translation of greatest treatise on technological chemistry, engineering, geology, mining of early modern times (1556). All 289 original woodcuts. 638pp. $6^3/_4$ x 11. 0-486-60006-8

FUNDAMENTALS OF ASTRODYNAMICS, Roger Bate et al. Modern approach developed by U.S. Air Force Academy. Designed as a first course. Problems, exercises. Numerous illustrations. 455pp. $5^3/_8$ x $8^1/_2$. 0-486-60061-0

DYNAMICS OF FLUIDS IN POROUS MEDIA, Jacob Bear. For advanced students of ground water hydrology, soil mechanics and physics, drainage and irrigation engineering and more. 335 illustrations. Exercises, with answers. 784pp. $6^1/_8$ x $9^1/_4$. 0-486-65675-6

THEORY OF VISCOELASTICITY (SECOND EDITION), Richard M. Christensen. Complete consistent description of the linear theory of the viscoelastic behavior of materials. Problem-solving techniques discussed. 1982 edition. 29 figures. xiv+364pp. $6^1/_8$ x $9^1/_4$. 0-486-42880-X

MECHANICS, J. P. Den Hartog. A classic introductory text or refresher. Hundreds of applications and design problems illuminate fundamentals of trusses, loaded beams and cables, etc. 334 answered problems. 462pp. $5^3/_8$ x $8^1/_2$. 0-486-60754-2

MECHANICAL VIBRATIONS, J. P. Den Hartog. Classic textbook offers lucid explanations and illustrative models, applying theories of vibrations to a variety of practical industrial engineering problems. Numerous figures. 233 problems, solutions. Appendix. Index. Preface. 436pp. $5^3/_8$ x $8^1/_2$. 0-486-64785-4

STRENGTH OF MATERIALS, J. P. Den Hartog. Full, clear treatment of basic material (tension, torsion, bending, etc.) plus advanced material on engineering methods, applications. 350 answered problems. 323pp. $5^3/_8$ x $8^1/_2$. 0-486-60755-0

A HISTORY OF MECHANICS, René Dugas. Monumental study of mechanical principles from antiquity to quantum mechanics. Contributions of ancient Greeks, Galileo, Leonardo, Kepler, Lagrange, many others. 671pp. $5^3/_8$ x $8^1/_2$. 0-486-65632-2

STABILITY THEORY AND ITS APPLICATIONS TO STRUCTURAL MECHANICS, Clive L. Dym. Self-contained text focuses on Koiter postbuckling analyses, with mathematical notions of stability of motion. Basing minimum energy principles for static stability upon dynamic concepts of stability of motion, it develops asymptotic buckling and postbuckling analyses from potential energy considerations, with applications to columns, plates, and arches. 1974 ed. 208pp. $5^3/_8$ x $8^1/_2$. 0-486-42541-X

BASIC ELECTRICITY, U.S. Bureau of Naval Personnel. Originally a training course; best nontechnical coverage. Topics include batteries, circuits, conductors, AC and DC, inductance and capacitance, generators, motors, transformers, amplifiers, etc. Many questions with answers. 349 illustrations. 1969 edition. 448pp. $6^1/_2$ x $9^1/_4$. 0-486-20973-3

ROCKETS, Robert Goddard. Two of the most significant publications in the history of rocketry and jet propulsion: "A Method of Reaching Extreme Altitudes" (1919) and "Liquid Propellant Rocket Development" (1936). 128pp. $5^3/_8$ x $8^1/_2$. 0-486-42537-1

STATISTICAL MECHANICS: PRINCIPLES AND APPLICATIONS, Terrell L. Hill. Standard text covers fundamentals of statistical mechanics, applications to fluctuation theory, imperfect gases, distribution functions, more. 448pp. $5^3/_8$ x $8^1/_2$. 0-486-65390-0

ENGINEERING AND TECHNOLOGY 1650–1750: ILLUSTRATIONS AND TEXTS FROM ORIGINAL SOURCES, Martin Jensen. Highly readable text with more than 200 contemporary drawings and detailed engravings of engineering projects dealing with surveying, leveling, materials, hand tools, lifting equipment, transport and erection, piling, bailing, water supply, hydraulic engineering, and more. Among the specific projects outlined-transporting a 50-ton stone to the Louvre, erecting an obelisk, building timber locks, and dredging canals. 207pp. $8^3/_8$ x $11^1/_4$. 0-486-42232-1

THE VARIATIONAL PRINCIPLES OF MECHANICS, Cornelius Lanczos. Graduate level coverage of calculus of variations, equations of motion, relativistic mechanics, more. First inexpensive paperbound edition of classic treatise. Index. Bibliography. 418pp. $5^3/_8$ x $8^1/_2$. 0-486-65067-7

PROTECTION OF ELECTRONIC CIRCUITS FROM OVERVOLTAGES, Ronald B. Standler. Five-part treatment presents practical rules and strategies for circuits designed to protect electronic systems from damage by transient overvoltages. 1989 ed. xxiv+434pp. $6^1/_8$ x $9^1/_4$. 0-486-42552-5

ROTARY WING AERODYNAMICS, W. Z. Stepniewski. Clear, concise text covers aerodynamic phenomena of the rotor and offers guidelines for helicopter performance evaluation. Originally prepared for NASA. 537 figures. 640pp. $6^1/_8$ x $9^1/_4$. 0-486-64647-5

INTRODUCTION TO SPACE DYNAMICS, William Tyrrell Thomson. Comprehensive, classic introduction to space-flight engineering for advanced undergraduate and graduate students. Includes vector algebra, kinematics, transformation of coordinates. Bibliography. Index. 352pp. $5^3/_8$ x $8^1/_2$. 0-486-65113-4

HISTORY OF STRENGTH OF MATERIALS, Stephen P. Timoshenko. Excellent historical survey of the strength of materials with many references to the theories of elasticity and structure. 245 figures. 452pp. $5^3/_8$ x $8^1/_2$. 0-486-61187-6

ANALYTICAL FRACTURE MECHANICS, David J. Unger. Self-contained text supplements standard fracture mechanics texts by focusing on analytical methods for determining crack-tip stress and strain fields. 336pp. $6^1/_8$ x $9^1/_4$. 0-486-41737-9

STATISTICAL MECHANICS OF ELASTICITY, J. H. Weiner. Advanced, self-contained treatment illustrates general principles and elastic behavior of solids. Part 1, based on classical mechanics, studies thermoelastic behavior of crystalline and polymeric solids. Part 2, based on quantum mechanics, focuses on interatomic force laws, behavior of solids, and thermally activated processes. For students of physics and chemistry and for polymer physicists. 1983 ed. 96 figures. 496pp. $5^3/_8$ x $8^1/_2$. 0-486-42260-7

Mathematics

FUNCTIONAL ANALYSIS (Second Corrected Edition), George Bachman and Lawrence Narici. Excellent treatment of subject geared toward students with background in linear algebra, advanced calculus, physics and engineering. Text covers introduction to inner-product spaces, normed, metric spaces, and topological spaces; complete orthonormal sets, the Hahn-Banach Theorem and its consequences, and many other related subjects. 1966 ed. 544pp. $6^1/_8$ x $9^1/_4$. 0-486-40251-7

DIFFERENTIAL MANIFOLDS, Antoni A. Kosinski. Introductory text for advanced undergraduates and graduate students presents systematic study of the topological structure of smooth manifolds, starting with elements of theory and concluding with method of surgery. 1993 edition. 288pp. $5^3/_8$ x $8^1/_2$. 0-486-46244-7

VECTOR AND TENSOR ANALYSIS WITH APPLICATIONS, A. I. Borisenko and I. E. Tarapov. Concise introduction. Worked-out problems, solutions, exercises. 257pp. $5^3/_8$ x $8^1/_4$. 0-486-63833-2

AN INTRODUCTION TO ORDINARY DIFFERENTIAL EQUATIONS, Earl A. Coddington. A thorough and systematic first course in elementary differential equations for undergraduates in mathematics and science, with many exercises and problems (with answers). Index. 304pp. $5^3/_8$ x $8^1/_2$. 0-486-65942-9

FOURIER SERIES AND ORTHOGONAL FUNCTIONS, Harry F. Davis. An incisive text combining theory and practical example to introduce Fourier series, orthogonal functions and applications of the Fourier method to boundary-value problems. 570 exercises. Answers and notes. 416pp. $5^3/_8$ x $8^1/_2$. 0-486-65973-9

COMPUTABILITY AND UNSOLVABILITY, Martin Davis. Classic graduate-level introduction to theory of computability, usually referred to as theory of recurrent functions. New preface and appendix. 288pp. $5^3/_8$ x $8^1/_2$. 0-486-61471-9

AN INTRODUCTION TO MATHEMATICAL ANALYSIS, Robert A. Rankin. Dealing chiefly with functions of a single real variable, this text by a distinguished educator introduces limits, continuity, differentiability, integration, convergence of infinite series, double series, and infinite products. 1963 edition. 624pp. $5^3/_8$ x $8^1/_2$. 0-486-46251-X

METHODS OF NUMERICAL INTEGRATION (SECOND EDITION), Philip J. Davis and Philip Rabinowitz. Requiring only a background in calculus, this text covers approximate integration over finite and infinite intervals, error analysis, approximate integration in two or more dimensions, and automatic integration. 1984 edition. 624pp. $5^3/_8$ x $8^1/_2$. 0-486-45339-1

INTRODUCTION TO LINEAR ALGEBRA AND DIFFERENTIAL EQUATIONS, John W. Dettman. Excellent text covers complex numbers, determinants, orthonormal bases, Laplace transforms, much more. Exercises with solutions. Undergraduate level. 416pp. $5^3/_8$ x $8^1/_2$. 0-486-65191-6

RIEMANN'S ZETA FUNCTION, H. M. Edwards. Superb, high-level study of landmark 1859 publication entitled "On the Number of Primes Less Than a Given Magnitude" traces developments in mathematical theory that it inspired. xiv+315pp. $5^3/_8$ x $8^1/_2$. 0-486-41740-9

CALCULUS OF VARIATIONS WITH APPLICATIONS, George M. Ewing. Applications-oriented introduction to variational theory develops insight and promotes understanding of specialized books, research papers. Suitable for advanced undergraduate/graduate students as primary, supplementary text. 352pp. $5^3/_8$ x $8^1/_2$. 0-486-64856-7

MATHEMATICIAN'S DELIGHT, W. W. Sawyer. "Recommended with confidence" by *The Times Literary Supplement,* this lively survey was written by a renowned teacher. It starts with arithmetic and algebra, gradually proceeding to trigonometry and calculus. 1943 edition. 240pp. $5^3/_8$ x $8^1/_2$. 0-486-46240-4

ADVANCED EUCLIDEAN GEOMETRY, Roger A. Johnson. This classic text explores the geometry of the triangle and the circle, concentrating on extensions of Euclidean theory, and examining in detail many relatively recent theorems. 1929 edition. 336pp. $5^3/_8$ x $8^1/_2$. 0-486-46237-4

COUNTEREXAMPLES IN ANALYSIS, Bernard R. Gelbaum and John M. H. Olmsted. These counterexamples deal mostly with the part of analysis known as "real variables." The first half covers the real number system, and the second half encompasses higher dimensions. 1962 edition. xxiv+198pp. $5^3/_8$ x $8^1/_2$. 0-486-42875-3

CATASTROPHE THEORY FOR SCIENTISTS AND ENGINEERS, Robert Gilmore. Advanced-level treatment describes mathematics of theory grounded in the work of Poincaré, R. Thom, other mathematicians. Also important applications to problems in mathematics, physics, chemistry and engineering. 1981 edition. References. 28 tables. 397 black-and-white illustrations. xvii + 666pp. $6^1/_8$ x $9^1/_4$. 0-486-67539-4

COMPLEX VARIABLES: Second Edition, Robert B. Ash and W. P. Novinger. Suitable for advanced undergraduates and graduate students, this newly revised treatment covers Cauchy theorem and its applications, analytic functions, and the prime number theorem. Numerous problems and solutions. 2004 edition. 224pp. $6^1/_2$ x $9^1/_4$. 0-486-46250-1

NUMERICAL METHODS FOR SCIENTISTS AND ENGINEERS, Richard Hamming. Classic text stresses frequency approach in coverage of algorithms, polynomial approximation, Fourier approximation, exponential approximation, other topics. Revised and enlarged 2nd edition. 721pp. $5^3/_8$ x $8^1/_2$. 0-486-65241-6

INTRODUCTION TO NUMERICAL ANALYSIS (2nd Edition), F. B. Hildebrand. Classic, fundamental treatment covers computation, approximation, interpolation, numerical differentiation and integration, other topics. 150 new problems. 669pp. $5^3/_8$ x $8^1/_2$. 0-486-65363-3

MARKOV PROCESSES AND POTENTIAL THEORY, Robert M. Blumental and Ronald K. Getoor. This graduate-level text explores the relationship between Markov processes and potential theory in terms of excessive functions, multiplicative functionals and subprocesses, additive functionals and their potentials, and dual processes. 1968 edition. 320pp. $5^3/_8$ x $8^1/_2$. 0-486-46263-3

ABSTRACT SETS AND FINITE ORDINALS: An Introduction to the Study of Set Theory, G. B. Keene. This text unites logical and philosophical aspects of set theory in a manner intelligible to mathematicians without training in formal logic and to logicians without a mathematical background. 1961 edition. 112pp. $5^3/_8$ x $8^1/_2$. 0-486-46249-8

CATALOG OF DOVER BOOKS

A TREATISE ON ELECTRICITY AND MAGNETISM, James Clerk Maxwell. Important foundation work of modern physics. Brings to final form Maxwell's theory of electromagnetism and rigorously derives his general equations of field theory. 1,084pp. $5^3/_8$ x $8^1/_2$. Two-vol. set. Vol. I: 0-486-60636-8 Vol. II: 0-486-60637-6

MATHEMATICS FOR PHYSICISTS, Philippe Dennery and Andre Krzywicki. Superb text provides math needed to understand today's more advanced topics in physics and engineering. Theory of functions of a complex variable, linear vector spaces, much more. Problems. 1967 edition. 400pp. $6^1/_2$ x $9^1/_4$. 0-486-69193-4

INTRODUCTION TO QUANTUM MECHANICS WITH APPLICATIONS TO CHEMISTRY, Linus Pauling & E. Bright Wilson, Jr. Classic undergraduate text by Nobel Prize winner applies quantum mechanics to chemical and physical problems. Numerous tables and figures enhance the text. Chapter bibliographies. Appendices. Index. 468pp. $5^3/_8$ x $8^1/_2$. 0-486-64871-0

METHODS OF THERMODYNAMICS, Howard Reiss. Outstanding text focuses on physical technique of thermodynamics, typical problem areas of understanding, and significance and use of thermodynamic potential. 1965 edition. 238pp. $5^3/_8$ x $8^1/_2$. 0-486-69445-3

THE ELECTROMAGNETIC FIELD, Albert Shadowitz. Comprehensive under- graduate text covers basics of electric and magnetic fields, builds up to electromagnetic theory. Also related topics, including relativity. Over 900 problems. 768pp. $5^5/_8$ x $8^1/_4$. 0-486-65660-8

GREAT EXPERIMENTS IN PHYSICS: FIRSTHAND ACCOUNTS FROM GALILEO TO EINSTEIN, Morris H. Shamos (ed.). 25 crucial discoveries: Newton's laws of motion, Chadwick's study of the neutron, Hertz on electromagnetic waves, more. Original accounts clearly annotated. 370pp. $5^3/_8$ x $8^1/_2$. 0-486-25346-5

EINSTEIN'S LEGACY, Julian Schwinger. A Nobel Laureate relates fascinating story of Einstein and development of relativity theory in well-illustrated, nontechnical volume. Subjects include meaning of time, paradoxes of space travel, gravity and its effect on light, non-Euclidean geometry and curving of space-time, impact of radio astronomy and space-age discoveries, and more. 189 b/w illustrations. xiv+250pp. $8^3/_8$ x $9^1/_4$. 0-486-41974-6

THE VARIATIONAL PRINCIPLES OF MECHANICS, Cornelius Lanczos. Philosophic, less formalistic approach to analytical mechanics offers model of clear, scholarly exposition at graduate level with coverage of basics, calculus of variations, principle of virtual work, equations of motion, more. 418pp. $5^3/_8$ x $8^1/_2$. 0-486-65067-7